Volunteer Training Officer's Handbook

Volunteer Training Officer's Handbook

W. Edward Buchanan, Jr.

Tulsa, Oklahoma

Copyright © 2003 by
PennWell Corporation
1421 South Sheridan
Tulsa, Oklahoma 74112 USA

800.752.9764
1.918.831.9421
sales@pennwell.com
www.pennwell-store.com
www.pennwell.com

Supervising Editor: Jared Wicklund
Production Editor: Sue Rhodes Dodd
Cover and book design: Clark Bell

Library of Congress Cataloging-in-Publication Data
Buchanan, W. Edward, 1966-
  Volunteer training officer's handbook / by W. Edward Buchanan, Jr..
     p. cm.
  ISBN 0-87814-834-5 (softcover)
  1. Fire fighters--Training of. 2. Fire extinction--Study and teaching.
3. Volunteer fire departments. I. Title.
  TH9120 .B83 2003
  628.9'25'071--dc21

                                                          2003003653

Printed in the United States of America

1 2 3 4 5 07 06 05 04 03

This book is dedicated to all those who gave their lives in the line of duty.

We'll miss you, brothers.

Contents

Preface

The volunteer fire service is extremely diverse, and each volunteer training officer or instructor has a unique set of challenges that they must face. This book presents an approach to volunteer training that is deeply rooted in the proud traditions of the fire service and is focused on creating volunteer firefighters who are professional and competent. The content is centered on the academy format for volunteer oriented training. At first glance, many volunteer training officers will assume they simply cannot keep pace with the academy format because the schedule seems too demanding. It is important to remember that the training approach presented revolves around a two-nights/week and every-other-Saturday format, a schedule very common to the volunteer training system. The academy format is simply a different approach to volunteer training, using an already familiar scheduling formula.

Each volunteer training officer must evaluate their department to determine whether or not they are prepared for such a program. The process is designed so that the training officer can use what meets the individual needs of their department and so that the principles can be implemented at a comfortable pace. Not all facets of the training system will work for every department, and the training officer must determine what is acceptable in their specific case. Using the training approach described in this book, the training officer can simply use those techniques right for them.

Acknowledgements

Many contributed to the creation of this book.
Without them, none of this would be possible.

Michael G. Harman, Sr.
Director, Hanover County (VA) Public Safety Department

Fred Crosby
Fire and EMS Chief, Hanover County (VA) Public Safety Department

Perry W. Hornbarger
Senior Captain, Chesterfield (VA) Fire Department

Rick Lasky
Chief, Lewisville (TX) Fire Department

Seth Dale
Instructor, Chief Shabbona (IL) Fire Academy

Rich Collins
Assistant Chief, Southwest United Fire Academy (IL)

Katherine Ridenhour
Captain, Aurora (CO) Fire Department

Martin LaRusso
Battalion Chief, Aurora (CO) Fire Department

The Hanover Fire Academy (VA) Staff
Ladd Grindstaff
Jason Kerrick
Greg Martin
Wanda Harris
Gene Hall

Adam Thiel
Executive Director, Virginia Department of Fire Programs

Larry McAndrews
Virginia Department of Fire Programs

The FDIC firefighter survival gang—The brothers!

Thanks to Tori and Jake for putting up with me during this process!

1

Self-Preparation

There is no greater honor than to educate the volunteer fire service. Today, children buy firefighter costumes for Halloween and ask for firefighter action figures at Christmas and for birthdays. Firefighters are becoming known for the heroic actions that they themselves have always viewed as just part of the job. The volunteer training officer's job is to train America's heroes and there should be great pride in this task and responsibility. The role as volunteer training officer will be challenging and rewarding, with a wide variety of highs and lows, and most will be placed in situations not faced by other members of the fire service. New training officers will be tasked with finding the balance between change and tradition as the fire service evolves into the future. Some issues will bring praise from friends and coworkers and others will bring discomfort, forcing the department to negotiate stressful changes. Either way, training officers play a vital role in sculpting every department's future. How the individual meets these challenges will determine their training legacy, whether it is over a short, several-year period or a lifelong career in training.

Each volunteer training officer will have a set of challenges that is specific to their jurisdiction or locality. Volunteer training programs can often be subject to internal political scrutiny, and what works for one department may not work for another. Despite these differences, many of the challenges faced by today's volunteer fire departments are similar; all face changes in society that can stretch the abilities of the average volunteer to their limits. The theories proposed in this book are designed to offer the training officer ideas to keep their training programs focused on quality and efficiency, rather than the political clutter that sometimes finds its way into the classroom. Not all theories will work for every training officer, but many departments have found success using these leadership and management principles. Each volunteer training officer must evaluate their department to determine what is feasible based on their current level of development and available resources. This chapter is designed to have the volunteer training officers examine themselves first, to ensure they are ready to lead their department into the future.

Professional vs. Volunteer

Sadly, the fire service has informally divided itself into two groups, those who receive a paycheck for their efforts, and those who do not. It is common for the group receiving monetary compensation to be referred to as professional fire-fighters, while those who do not are referred to as volunteer firefighters. Classifying the fire service in this manner creates a false impression that the service provided by volunteers is somehow less valuable than the paid service. This false impression is generated in a number of ways. Paid firefighters tend to feel a sense of superiority over the volunteer force, based simply on the fact that they get paid; yet, many of them got their start as volunteers. Volunteer firefighters can sometimes create a false impression themselves by adopting a "We can't, we're just volunteers" attitude to disguise a lack of confidence or downright laziness. Experience shows that some volunteer firefighters are very professional, becoming highly educated and providing excellent service to their communities. On the other hand, firefighters receiving paychecks sometimes lack the education and ability to meet the high standards of excellence passed on through the history of the fire service.

The volunteer training officer must understand the differences between the paid and volunteer sectors, as they must be able to tailor their training programs to accommodate specific volunteer needs. To help identify these contrasts, the definition of profession must be examined. The International City Management Association describes a profession as having the following four basic elements: (1) a broad base of knowledge, such as modern fire science research in material burn characteristics and suppression techniques, (2) competent application, such as qualification systems like the National Professional Qualifications Board (NPQB), (3) social responsibility as evidenced in enhanced fire prevention programs and hazardous-materials control, and (4) self-regulation, such as the National Fire Protection Association (NFPA), which creates professional standards for the fire service.[1] This description outlines the fire service as a profession in general. The sep-aration between paid and volunteer in relationship to professionalism was derived internally to the fire service over time, without consideration of actual ability.

The broad scope of the paid and volunteer fire services are similar; both provide services based on the needs of the locality (Fig. 1–1). The key differences between the paid and volunteer services are funding resources and retention practices. Paid departments rely heavily on budget allotments from the local tax base to fund their operations. Policies sometimes prevent local government from soliciting funding through fund drives and various fund-raising events. The budget allotments can be millions of dollars, but when the money runs out, there is no further recourse, barring the rare grant opportunity. Volunteer departments may or may not receive funding from the local tax base, but have the advantage, as nonprofit organizations, to conduct fund drives and other fundraisers to gain the money needed for providing services. A wide variety of options are available to raise money from fill-the-boot drives to bingo.

Paid Volunteer

Career

- Provides fire and EMS services 24 hours a day
- Attends required training in a variety of disciplines
- Conveys a sense of professionalism at all times
- Receives funding through local tax base and may apply for grants
- Enjoys consistent retention statistics due to career orientation

Volunteer

- Provides fire and EMS services 24 hours a day
- Attends required training in a variety of disciplines
- Conveys a sense of professionalism at all times
- May receive funding through fund drives, local tax base, and/or grants
- Limited success in retention

Fig. 1–1 Paid/Volunteer Training Officer Responsibilities (EMS—emergency medical services)

Volunteer departments experience more challenges in retaining personnel than paid departments. Paid firefighters enjoy schedules that allow them to supplement their income by working second jobs, thus providing a direct benefit to their families. Volunteer firefighters have full-time jobs away from the fire service and volunteerism can have a negative impact on the family, causing them to spend hours away from home with little or no positive financial impact.

Despite the specific differences between the paid and volunteer departments, the service mission of the two remains the same. Therefore, differences between paid and volunteer services that must be addressed by the volunteer training officer lie in the delivery systems, not the final outcomes. Both paid and volunteer training officers must remain focused on creating the end product, which from the citizen's perspective, is the traditional professional firefighter.

The majority of the delivery systems discussed in later chapters focus on the concept of the volunteer fire academy. It is import to realize that the academy format is dynamic and very flexible. You as the training officer can use as much or as little of this program as will fit within your department's culture and political environment. There is no doubt that with change comes controversy, and the departments used as examples were no exception. But with focus, perseverance, and thoughtful implementation, the resistance to change can be overcome. In the volunteer academy approach, the only true difference between the volunteer program and what some career departments might do is the schedule. Remember that your job as training officer is to create firefighters. The NFPA finds no difference in these qualifications between career and volunteer, and neither should we. We must be as creative as possible in the scheduling and delivery, but when it comes to content, we must stand fast—for the safety of our citizens and ourselves.

The Training Officer and the Job

Many times in the volunteer fire service, the training officer position, along with other leadership positions, is appointed or elected. While this system is democratic in nature, it sometimes lacks accuracy in placing qualified personnel in key positions. Occasionally, the training position is filled, not by appointing the most qualified person, but by accepting someone who is willing to take on the job. This situation often finds a well-meaning person trying to do the best they can, while receiving little information about the job from their predecessor. So, like it or not, you are suddenly elected or appointed the training officer for your volunteer department. Where do you start?

Good business principles teach us that for every job position there should be a job description listing the responsibilities of each component of the organization. Some of the job duties of the volunteer training officer include the following:

- Providing recruit training to new members
- Providing in-service training for current members

- Providing advanced training, such as technical rescue and hazardous materials
- Developing programs to meet new standards and requirements
- Monitoring the industry for new developments and techniques

(See sample job description on the CD-ROM)

To meet this list of requirements, one must possess knowledge, skills, and abilities beyond those of the regular volunteer firefighter. The training officer will be the individual that the entire department looks to for answers and guidance in the future. Even the fire chief turns to training personnel for answers about new and technical aspects of the fire service. Yet, to truly capture the essence of the job, we must look beyond the formal job description. The real mission will one day be reflected in the ability of the volunteers trained by the system.

The decision

Every firefighter's fate, paid or volunteer, rests on a decision they may one day be forced to make when they find themselves in that hot, smoky hallway searching for a missing child, and the conditions suddenly change for the worst. They have to make a decision to stay or go based on their experience, knowledge, and skills. "Do I keep searching and maybe rescue the child, or get out now?" And what are they wagering on this decision? They risk never seeing their family and friends again, leaving children fatherless or motherless, and possibly dying a horrible fiery death. But what if they opt to bail out and are left wondering, "Maybe if I had pushed a little further?" There will always be cases of *what if,* but they can be greatly reduced by creating highly educated and skilled firefighters, confident in their abilities and aware of the risks. This is the root of the training officer's mission. Firefighters must be provided with a knowledge base that will serve them well when their time comes to make that decision, "Do I stay and push on, or get out now?"

Understanding the true mission of the volunteer training officer or instructor is vital to ensuring future success. A passionate fire service instructor created this instructor creed to be posted and distributed to the various instructors at the Hanover (VA) Fire Academy.

Instructor Creed

I am a Fire Service Instructor, guardian of the traditions and knowledge of my brethren before me. No one is more professional than I. Competence is my watchword and proficiency is my shield. I am the source of the knowledge and skills to be passed on to the future of our profession. I shall never settle for mediocrity, while excellence is my goal. Integrity is the lifeblood of my calling and I shall never cause it to be brought into question. My actions and deeds shall be the measure by which I am judged. Two things shall remain foremost in my mind, accomplishing my mission and the safety of my charges. They cannot do their job if I have failed in mine, for I am a Fire Service Instructor.

—Thomas Nelson
Hanover (VA) Fire Academy

This creed may be placed in various locations to remind the instructor and training officer of their mission on a daily basis. This message may be printed on wallet-sized cards so that they may never forget their responsibilities.

Though the training officer position requires the highest level of dedication, it also offers the highest rewards. The training staff is often chosen to attend national conferences and meetings because they can bring back valuable information and distribute it to the department. They are also privileged to unique experiences in the fire service, such as serving on interview panels, or special committees, etc. These opportunities often prove beneficial years later as they progress through the leadership ranks.

The first task toward becoming a training officer is to read, study, and learn as much as possible about our great fire service. Be current on the latest rescue methods, the most efficient pump calculations, and anything that might impact the job. Regardless of their personal training background, training officers can read the trade journals (not just looking at the pictures) to keep up with the latest information. This immediate change requires very few resources. The training officer must learn the information and pass it on.

The training officer and instructor must forever be a student of the fire service. To be confident in front of a classroom, one must be confident in their abilities on the fireground. Our brothers and sisters in the fire service will immediately spot those not mentally and physically prepared. This is not a change that can take place overnight, but it can be managed with time and patience. Simply take every class possible! Find ways to hone and sharpen your abilities to make yourself the best you can be. Community colleges often provide associate degrees in fire science technology, or simply classes in building construction or fire behavior. Many people started out taking just a class or two, and wound up graduating from the program. Beware of those who believe they have seen it all, for they will soon stumble. No matter your experience, you can ready yourself for the task. Constantly try to learn new things and it will benefit you and your department for years to come.

Mastering Your Traditions to Promote Change

Many years ago when the fire service relied on horses and wagons to deliver fire suppression, it was common for the firefighters to wipe down the wheels upon returning to the station. Because the spokes of the wheels were made of wood, they were carefully maintained to ensure reliable performance. Not so long ago, two firefighters were wiping down the wheels of a modern fire apparatus when they realized that there didn't seem to be a logical reason for this procedure. They asked their lieutenant why this task is performed, but he didn't know either. Their standard operating procedures specifically indicated that the wheels must be wiped down upon returning to the station, yet no one seemed to know exactly why. In the tiny print at the bottom of the procedure, they noticed the revision date—July 1, 1922. The "wipe-the-wheels" policy had simply been left on the books, in spite of technological advances in fire apparatus. Once again, the fire service had carried on a tradition, not because it was practical, but because it was tradition.

Much like the firefighters in the story, mastering traditions will promote change. As departments move into the future, the training officer must pass the rich history of the fire service on to the next generation. Only by knowing where we have been can we truly plot a course for the future. Otherwise, we will just show up in *tomorrow* by accident, and it may not be a location of our choosing.

Share with the volunteer recruits stories of how the pike pole came to be. Tell them how the early settlers of our country used them to pull down houses made of straw and mud to create a firebreak. What many refer to as bugles are actually speaking trumpets used to amplify verbal commands on the fireground. Tell them why we are the way we are. It is much easier to accept change when you have a full understanding of fire service evolution. The volunteer training officer must be the voice of reason, bringing logic and hard data to sometimes emotionally charged issues. The key to overcoming resistance to change is through education and training.

Serving as a department's agent for change, some issues will be championed and some will not; yet, the training officer must find ways to facilitate these transitions. How you implement change is key to your success. Former friends may be very displeased as new training requirements are introduced. You may not survive if you attempt to implement change inappropriately. Failing to implement change carefully can be a fatal mistake for rookie training officers.

The up side to the change issue is that most volunteer departments share in the same struggle. How do we meet growing training demands and call loads with what seems to be a dwindling volunteer force? The solution to this universal problem will require both training officers and volunteer firefighters to step back from the conventional programs of the past and be open to new training ideas. The solutions will not be easy. The training requirements for 21st century firefighters will not get easier or go away. We cannot afford to grow lax in our training standards just because it is difficult. Our only option is to alter the delivery system so that training is as convenient as possible for those who are selected to volunteer. It is vital that we educate our volunteers to ensure they are instilled with a sense of pride and love for the fire service that our predecessors deserve. Many training programs can deliver the technical skills required to be a firefighter, but there is much more to being a firefighter than just skill. With being a firefighter comes expectations from both the public and the fire service. As new volunteers enter the fire service they join a vast family, a family with expectations and family values. How will the new volunteer survive in this new family if they do not understand the expectations?

Fire Service Training Philosophies

There are several principles that serve as the cornerstones of effective volunteer training programs. These principles must be understood to effectively pass them on to the student. Instructors serve as an example of these fire service principles and they should ensure that the principles are passed on, as the survival of the volunteer fire service depends on it. These cornerstones are honor, discipline, responsibility, and accountability.

Honor

Honor essentially means having a sense of ethical conduct, or being as good as your word and doing the right thing. The fire service translates honor as the glue that binds brother and sister together in helping their neighbor, regardless of social class, living conditions, or geography. This translation implies that honor must be demonstrated on the following two fronts: (1) within the fire service itself, and (2) in the service we provide the public.

Inside the fire service, firefighters must understand that the term *brotherhood* is not some cliché or membership in a social club. It means that on duty or off duty, rain or shine, firefighters are there to help their fellow citizens under any circumstances, no matter the potential risk or inconvenience. A true brother and progressive leader of the fire service, Chief Rick Lasky of the Lewisville, Texas Fire Department said it best,

> *The brotherhood to me means more than just a sticker on the windshield of your car. It means that when your kids are sick, we help out. That when you're having a tough time with your bills, we help out. That when you need to move into your new house, we move you; and when that house needs a new roof, we roof it. It also means that I would lie next to you and burn the ears off my head before I would ever leave you in a burning building.*[2]

The training officer must be prepared to help a brother no matter what the costs.

Over the last decade, the fire service has embraced the mission statement as a way to publicly state what they offer their community. Some departments require every firefighter to memorize this slogan as a way to unify their department. As training officers, we may be better served by creating an "honor statement" in conjunction with a mission statement. Let's face it; the public has a good understanding of what the fire department does. Ask new volunteers how honor plays a role in the fire service, and you'll likely get some blank looks. They will not understand unless we explain what honor means to the fire service.

Creating an honor statement simply requires outlining what is important to each of us in the volunteer fire service. What promises will you make to your brother or sister? How will you be there for him or her? This should be memorized and carried in their hearts forever. By establishing this relationship up front, we build a solid foundation that will serve us well for years to come. When fire service training is hung on a framework of history and tradition, it has double the impact on the student.

Sample honor statement: *Brother first—Duty always*

This sample honor statement is simple, yet it addresses the core values of the fire service. No matter what, we must take care of each other and we will never forget the honor of serving our communities. Protecting our brother and neighbor is an awesome responsibility. The honor statement helps solidify this distinction in both the instructor and the student.

Outside the fire service, we owe it to our communities to do the job in an honorable way. Look at the way second graders look at you next time you do a prevention program. No matter whether you have earned the title or not, you are a hero in their eyes, simply because you wear the uniform. The uniform itself is a symbol of honor. Wearing the uniform incurs an extra responsibility, one of longstanding tradition and honor. Training officers must act honorably on and off duty. We owe this to our other brothers and sisters to set an honorable example.

In recent times, law enforcement has asked to use fire department equipment and uniforms to infiltrate illegal operations to capitalize on the "We are here to help" reputation of the fire department. This is of great concern to fire service leaders because it may cause the public to question our honor or intentions. There is an unspoken, unwritten code between the fire service and their communities. Firefighters will help you without judgment or prejudice, regardless of your beliefs, appearance, or living conditions; all you need to do is call. That is an honor code we should treasure and protect. Ensure that our new volunteers are up to maintaining this code by making honor an ongoing component of the fire training program.

Discipline

It is no secret that firefighting is dangerous business. On the fireground, we are required to act with focus and accuracy in the most unimaginable situations. To survive, we must display steadfast discipline at all times. Unfortunately, this is a fading quality among our young volunteer recruits. The training officer must be a shining example of discipline, focus, and control. There is an old saying in the volunteer fire service, "You volunteer twice in the fire service, once to get in and once to get out. Everything else is mandatory." Though this mindset does not do much to enhance volunteer retention efforts, it does have a ring of truth in it. Firefighters can be seriously hurt or killed if they fail to follow orders on the fireground. Our young students must understand that fire service strategies and tactics are not created through a democratic process. They are based on strategic and tactical considerations derived through the rapid assessment of situations and resources.

Converting a new volunteer, whom society has taught to run from burning buildings, into a trained, measured, and accurate firefighting force requires a great deal of focus on discipline. Volunteers will emulate the amount of discipline and control they see in their department leaders. Whether in front of the classroom on Saturday morning or sitting on the tailboard on Saturday night, your volunteers will notice how disciplined others are around them. The training officer is the example they will follow, on or off duty.

Responsibility

The common definition for responsibility usually implies a sense of being limited or restricted by a task or obligation. For this application, listen to how the word sounds, *response ability*. Each of us has the ability to choose our responses and make decisions. Instead of feeling limited or restricted by this term, feel empowered. Each of us has the authority to dictate our personal responses to all of life's situations. When life is examined at its most basic levels, our situations are derived based on our actions and how we adapt to various occurrences in our lives. We inevitably have a right to choose, whether we choose to acknowledge it or not. A common statement (or joke) around the volunteer firehouse is, "I would have attended the training, but I had to cut the grass." When the response-ability theory is applied, we find they did not *have* to cut the grass, instead, they chose to do so based on their priorities at the time. Not to say there is dishonor in cutting the grass, because that certainly must be done, but understand that we have the ability to make choices and should be responsible for the decisions when we make them.

As training officers, we have the responsibility to promote training, not as something we *have* to do, but as a task we have chosen because it is positive. We have two choices when confronted with a new or extended training requirement. We can gripe about it, conveying an inherent sense of negativity amongst our firefighters, or we can present the need for the new training and why it is important to our service delivery. Negativity is highly contagious, especially when coming from the volunteer training officer. Remember, volunteers will be looking to the training officer to see how they have embraced the new requirement and will form their impressions based on their trainer's response.

This same positive presentation is vital in all facets of the fire service. Yes, there will be those pockets of folks who will gather behind the engine to gossip and speak negatively about nearly everyone in your fire company, but do not be intimidated or coaxed into participation. The root of their negativity usually stems from either ignorance of the facts or insecurity in their abilities. The training officer's job will be to present the facts clearly and promote teamwork and cooperation. If they are ignorant, the training officer must educate them. If they are insecure, work with them one-on-one, so that they can be confident in themselves. There is no one else to take the job; you have the response-ability.

Accountability

Accountability, or a willingness to accept responsibility, has been a buzzword in the fire service for some time. When the incident command and management systems became popular in the 1980s, one of the proposed benefits was enhanced personnel accountability. Today, we have tags that Velcro®, tags that clip, and even some that are wireless transmitters. Evidently, the accountability offered from the incident management classes missed its mark if we need these elaborate systems today.

By definition, our inability to account for firefighters on the fireground is a direct result of our unwillingness to accept responsibility for our own actions. This inability or unwillingness to accept responsibility for our actions debilitates us not only on the fireground, but also in the fire station where most of our volunteer retention problems begin. The training officer must introduce volunteers to accountability on all levels. Those unwilling to be accountable for their actions must be weaned from the ranks, even amidst the recruitment challenges of today. Those who refuse to be accountable will slowly tear the department apart from the inside. Yes, we need volunteers, but there is a limit to who can make the grade.

Complaints are common among the more experienced volunteers about the new generation of volunteers coming into our ranks. They are believed to lack the moral integrity the old-timers surely possessed when they joined the department. Do we hold society accountable for creating such a lethargic crop of recruits? Absolutely not. It is our job to create firefighters who measure up to the high standards of our history. If there is anyone who should be held accountable, it is the very people who complain. Veteran firefighters have a responsibility to share the experience they have gained over the years. What was important to firefighters 50 years ago is fundamentally the same for firefighters today. If the volunteers we recruit today do not share in the ideals of our past, then it is up to those who know to pass the information on to the next generation. This same theme applies across the fire service. Accountability really has nothing to do with tags or icons. It is more about demonstrating honor, discipline, and responsibility. It is the tool by which we measure ourselves against the high moral standards expected of us by our peers and our communities. We must hold each other accountable for our actions. Our fire service is exactly what we make of it. Our actions as instructors in the classroom and on the drill ground lay the foundation for our future. We must ensure the next generation has the same love of the job that we do. Our forefathers are watching and history will record our performance.

Networking and Mentoring

The responsibilities as volunteer training officer are huge and the tasks sometimes overwhelming. Look across the country and you will find that many volunteer fire departments face similar challenges. Networking to share information is crucial to creating a successful volunteer training program. The training officer must be *in the loop* to fully realize the potential of their jurisdiction or regional area. The Southwest United Fire Academy (SUFA) in Darien, Illinois is a shining example of how networking to combine resources enhances the training officer's ability to deliver high-quality programs that would be unobtainable alone. The SUFA is a combination of four smaller departments that together provide a full range of fire service training. Each jurisdiction involved provides physical resources, such as apparatus and hand tools, plus instructors to produce volunteer fire academies that exemplify all of the training philosophies we have discussed in this chapter. Volunteer training officers can follow this example by contacting neighboring jurisdictions to discuss training needs. By sharing resources, you can manage training programs much larger in scale than could be done by one person in one department.

The instructor should be exposed to as many different people in like positions as possible. By attending meetings, conferences, and conventions in their area, training officers will meet instructors like themselves who are more than happy to share information. There are often meetings at the state and local levels that are inexpensive and require little time away from the family to attend. If the budget will support a larger trip, the FDIC is the conference to beat all others. Not only can instructors find training to enhance their personal skills as instructors, they can also meet thousands of people who share the same concerns. The people who attend the FDIC are there because they love the job and are happy to share ideas. Some instructors take vacation time and pay their own way to attend this exceptional event. It is highly recommended by everyone.

Mentoring

As the older members of our fire service retire from duty, so will their knowledge and experience unless it is passed on to the next generation. Some of the older members may not wish to offer their experiences unsolicited. It is up to the new leaders to ask the questions and listen to what the older members have to say. Ask them about the buildings they worry about burning in their area. Have long discussions about fire behavior and building construction.

Some of the older members may speak of that *career fire*, meaning the one defining moment of their entire fire service experience. Find out what happened and why. Capture their experiences and learn their lessons. For specific guidance as an instructor, seek out a mentor who has served in your position in the past. Climb under their wing and learn all you can. Let them serve as an advisor on not only the tough training issues, but on how to truly be a brother or a sister in this fire service. Ask them the questions you always wondered about, but were afraid to ask.

Training officers may not find a mentor in their department, or even their region. It may be someone they meet from across the country. Wherever they are found, stay in touch and share information. Learn all you can while you can. Before you realize it, someone will be looking to you for guidance. Pay your mentor back for their guidance by passing the lessons you learned on to the next generation.

Chapter Review Questions

1. List the four basic elements of a profession, as listed by the International City Management Association, and give an example of each.

2. What advantage do volunteer departments have over paid departments regarding funding?

3. List at least three of the volunteer training officer's basic job duties.

4. What are the four cornerstones of the fire service training philosophy?

5. What is the fire service definition of honor?

6. When incorporating change in the fire service, what is the training officer's best option, changing training requirements or altering the delivery system? Why?

Answers

1. Broad base of knowledge (modern fire science research), competent application (NPQB Board), social responsibility (fire prevention programs), and self-regulation (NFPA Standards).

2. Volunteer departments are free to conduct fundraisers as nonprofit organizations. Paid departments are often prevented from conducting such events without forming a separate nonprofit group. This allows volunteer departments to raise money without the limitations and bureaucracy associated with paid departments.

3. New recruit training, in-service training, advanced training, such as haz-mat or technical rescue, developing new programs, monitoring industry for new technology.

4. Honor, discipline, responsibility, accountability.

5. The glue that binds brother and sister together to help their neighbor, regardless of social class, living conditions, or geography.

6. The training officer's best option for incorporating change in the fire service is to alter the training delivery system rather than requirements, as the education requirements in high-hazard job functions are static, leaving the delivery system itself as the only fluid component.

Notes

[1] Marlatt, F. Patrick and Bruce J. Walz, 1988, *Managing Fire Services. 2nd. Ed.* Washington, D.C.: International City Management Association, 445–446

[2] Excerpt from Rick Lasky's Keynote Speech "Pride & Ownership," Fire Department Instructors Conference, 2002.

2

Assessing Department Needs

Deciding where to start is one of the most difficult challenges for the new volunteer training officer. Rather than choosing random goals and objectives, conduct a needs assessment to ensure the training mission reflects the true needs of the department. The new training officer will find a variety of members trying to exert influence over the training program so that it meets their individual needs. Unfortunately, the training needs of each volunteer member are vast, and attempting to meet such needs can send the training program into a tailspin. The volunteer training officer must learn to consider such input and weigh it against the true needs of the department through appropriate assessment and logic.

When replacing a previous volunteer training officer, determine what has worked in the past and any goals and objectives that may still be ongoing. The incumbent training officer should try to identify any training initiatives in the previous short- or long-term plan that may remain departmental priorities. Making a seamless transition from one training officer to the next should be an initial priority to ensure that the educational and training process continues. Once responsibilities have been passed to the new training officer, an assessment of current conditions can be conducted. Revised goals and objectives may then be established based on fact and good judgment rather than the popular trend of the month.

For smaller departments, the training officer transition may involve only a few people. For larger departments, the change can affect many individuals, and the change often has a much broader scope. The training officer in a large department must be prepared to make the transition from the front of the classroom to a management role. Organizing and allocating resources to meet the training mission will become the major job function. It is impossible to be involved in everything associated with an expansive training system and the training officer must be prepared to delegate responsibility appropriately. In most cases, this involves an instructor stepping up to the training officer position, which makes the transition more bearable because the new training officer has prior service in the training program. Being prepared to delegate will help the training officer absorb the huge responsibility of training a large department. This is an unnatural transition for some, but it must be mastered to survive in the modern volunteer fire service.

Performing a Needs Assessment

Consulting previous training officers is the first step in performing a needs assessment. It is important to understand the history and behind-the-scenes information before implementing change. Fact must be separated from opinion and fiction during this process. Avoid basing actions on statements such as, "This is how we've always done it" or because someone may be offended by changing a process. Change is seldom easy, but neither is progress. The outgoing training officer may be naturally defensive as you inquire about the process, while others may be very open. Be sensitive to their situation and be respectful of their abilities. Look for the facts surrounding the training program rather than opinions. Though opinions are important, you don't want them to taint your assessment. This is not to contradict the mentoring philosophies discussed in chapter 1, but rather being mindful of the facts while being respectful to others. This is a common early mistake for some volunteer training officers. Studying the facts and performance data in a sensitive manner will help focus training efforts on definitive goals, rather than on perception and opinion.

Since there are many stakeholders in the training system, an objective needs assessment should be conducted to identify the future direction of the training program. Three areas to consider when evaluating a department's present and future needs include the following: (1) organizational needs, (2) learner needs, and (3) job needs (Fig. 2–1). Each group of needs contains categories that influence the training program and should be considered in the needs assessment.

Organizational Needs	Learner Needs	Job Needs
Departmental goals	Perceived needs ■ Individual	Job analysis ■ Duties and tasks
Budget restraints	■ Group Actual needs ■ Maslow's hierarchy of needs ■ Herzberg's two-factor theory	■ Environment ■ Relationships ■ Requirements

Fig. 2–1 Needs Assessment Matrix

Organizational Needs

It is important to understand that the personal views of the training officer may not reflect that of the senior leadership in the department. The department leadership is responsible for determining the future plans through short- and long-range planning. Through these plans, goals and objectives are identified and relayed to the training officer. The competent and professional fire officer must be prepared to accept these goals as their own and work to see them made a reality. The training officer must remember that their job is to do what is best for the department, even when it is not consistent with their personal views. Those who become combative toward organizational needs tend to have very short careers as training officers.

Departmental goals

To determine the departmental goals, go directly to the departmental leaders via the chain of command. Ask to speak directly to the governing body to determine your immediate direction. This will provide you some guidance and buy some time to conduct a thorough needs assessment. Organizational needs should be consistent with the orientation provided by the outgoing training officer. Vast inconsistencies may indicate why the position has become vacant. Often, the organizational leaders will have several broad goals they would like to see accomplished. This may involve training initiatives that involve the whole department, such as survival training for everyone or increased pump operations training. Such initiatives must be factored into the needs assessment and be reflected in the subsequent goals and objectives. These goals often serve as the training skeleton for the training officer's first year on the job.

Budget restraints

Volunteer departments will often have lofty aspirations for the training program, but find their goals out of reach due to budget restraints. This presents a challenge to the new volunteer training officer to look for ways to meet their leaders' goals with limited resources. Broad training initiatives are often quite expensive and budgetary impacts go far beyond the initial cost of the training. Changes in equipment often go hand-in-hand with such initiatives and typically cost money. Look for initiatives that can be mastered with little impact on the training system and operations; those are easy wins for the new training officer. It will take time to become familiar with the job before any advanced budgetary magic should be attempted.

A perk of being a training officer is having the unique opportunity to influence organizational needs simply by having increased access to departmental leaders. Training officers often find themselves *go-to* problem solvers who have a unique influence on the department. Becoming an influential leader in the department will not be automatic. This status may take time to develop and requires the training officer to demonstrate sound decision-making abilities on a consistent basis. Realize that as training officer, you are constantly being evaluated by the senior leadership. Consistent and creative leadership and constant study will help transform the training officer into an influential member of the organization's leadership team.

Learner Needs

Whether in small groups or one-on-one, everybody has an opinion on what training is needed. Learner needs describe the volunteer firefighter's beliefs about what they need or want from the training program. Such learner needs can be categorized into *perceived* needs and *actual* needs. Each impacts the training program and must be addressed by the volunteer training officer. Failure to recognize either can result in unnecessary political upheavals that only make the training job more difficult.

Perceived learner needs

Perceived learner needs are normally based on the volunteer's current knowledge and skill base, seasoned with the member's perception of what is currently important. This perception may be accurate in some cases and not in others. The training officer must wade through these opinions to determine what best suits both the individual and the department as a whole.

When polling a group of volunteers, perceived learner needs tend to be responsive to recent incidents that have challenged the department. A serious vehicle accident that tested the department's abilities may cause the group to call for more vehicle extrication training, while companies responding to a drowning call are likely to request water rescue training. Unfortunately, these group-needs tend to be reactive rather than proactive and change regularly with the variety of calls run by the modern volunteer department. This makes it difficult for the training officer to meet individual needs as the landscape is constantly changing; not to mention, training efforts will always be after the fact. Through appropriate needs assessment, the training officer can identify risks for which the department currently lacks training and provide classes prior to an incident.

When polling volunteers one-on-one, perceived learner needs may reflect underlying agendas. The classic example for the volunteer fire service comes with each election of officers. This annual ritual causes predictable spikes in learner needs for training tailored to the requirements to serve in elected positions. Those motivated to serve in a leadership role are likely to aggressively seek out the requirements for a specific position. This perceived need for training is a direct reflection of the volunteer's personal agenda. Though each individual's needs are important, they shouldn't drive the training mission of the department. There will be as many different training agendas as there are firefighters.

Actual learner needs

Actual learner needs are related to the psychology of learning and motivation, principles often overlooked by the volunteer training officer. The training officer must understand learner motivations to get the most out of the training program without overtaxing both people and resources. Abraham Maslow and Charles Herzberg both developed theories about the needs and motivations of people. Each theory addresses different phases of training and both have merit when applied to the fire service.

Maslow's theory of motivation and needs

In 1943, psychologist Abraham Maslow proposed the theory that people are motivated based on a variety of needs, which Maslow categorized into five basic groups: physiological needs, safety needs, social needs, esteem needs, and self-actualization needs. These categories were then structured by the order of their importance, meaning that people tend to satisfy the most basic needs before moving on to the next level (Fig. 2–2).[1]

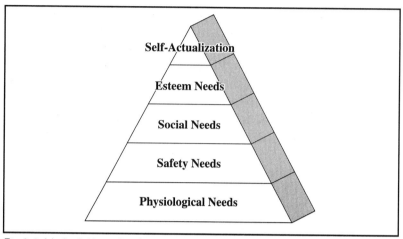

Fig. 2–2 Maslow's Hierarchy of Needs

The Hierarchy of Needs, as Maslow called it, has been used in a variety of applications, including psychology and business; it is even mentioned in the International Fire Service Training Association's (IFSTA's) Fire and Emergency Services Instructor Manual. Unfortunately, few in the fire service have taken the time to apply this theory in a way that demonstrates how it impacts the training mission on a daily basis. Figure 2–3 shows how each of the basic needs can be applied to fire training programs. The volunteer training officer will satisfy students' actual needs on a daily basis, in some cases, without even realizing it.

Physiological	**Safety**	**Social**	**Esteem**	**Self-Actualization**
Restrooms	Controlled training environments	Create an identity for each class	Maintain positive attitude	High standards
Comfortable environment				Graduation ceremonies
	Safety officer	Assign small groups or companies	Foster mutual respect	
Food/water				Special class awards
Regular breaks	Emergency procedures			
			Private counseling	

Fig. 2–3 Maslow's Hierarchy of Needs in the Fire Service

Physiological needs include our most basic human requirements, such as food, shelter, and clothing. Regular breaks, nearby food and refreshments, and appropriate training environments are common techniques used by instructors to meet some of these basic needs. The old training phrase, "The head can only absorb so much as the tail can endure," refers to the student's physiological need for rest, food, and water. Experienced instructors know that making the student physically comfortable and meeting physiological needs is the first step in preparing the training environment. Ensure that there are comfortable seats, refreshments, and restrooms nearby. If not, make provisions in the class schedule to allow students time to compensate. Allow extra break time, even if the class runs longer than expected, to ensure that the time spent in class is quality time, rather than quantity time.

Safety needs refer to the student's inherent need to survive. Learning will be notably hampered when the student is in fear for their life. Instructors and senior firefighters must remember that what seems normal to them may frighten the new volunteer recruit. Care must be taken to ensure that students are introduced to firefighting tactics at a pace that addresses their basic safety needs and allows time for students to build confidence in the instructor staff. Failure to do so will cause a sharp drop in retention statistics and eventually hamper recruitment efforts. Simply taking the time to explain tactics and procedures in a way that the new volunteer can understand will do wonders to provide a sense of confidence and safety.

Social needs refer to how one believes others perceive them or how they fit within a group of people. Students want to fit in among their peers and training programs should be designed to facilitate this need. In the training philosophies presented in this text, the need for social acceptance is associated with team building and responsibility to others, guiding the student toward achievement and success in ways not previously identified in most volunteer training programs. When people strive toward a goal as a group, social acceptance is achieved by helping the team meet its objectives. By reinforcing accountability at every level in the training process, students begin to understand the concept of teamwork and gain the respect of their peers. There are no individual failures. Those who perform below acceptable standards demonstrate a failure on the part of the team and a breakdown in accountability. By making everyone in a class accountable to one another, gaining social acceptance means working hard to ensure that *everyone* performs to standard. Students soon develop their own informal mentoring and tutoring programs and develop relationships that will last a lifetime.

To capitalize on students' social needs, large classes, such as firefighter certification programs, are given an identity. For example, use the year with a number that indicates a running succession of courses such as Class of 2001–School #5. Numbering each class gives each group a sense of social recognition and enables them to feel that they are part of something special. Although students may be assigned to different stations after graduating from basic firefighter training, they will always remember going through class together and will keep that shared bond with their classmates.

To further enhance the social structure, small companies consisting of 3 to 5 people are assigned for the duration of the course. Using alpha rather than numeric designations removes the sense of any preexisting political affiliations. For example, a class of 30 students divided into six companies of 5 students would use the company assignments Alpha, Bravo, Charlie, Delta, Echo, and Foxtrot. Mixing up the class members by department affiliation will also help to further break down any political tensions among the group. Students quickly find that even firefighters from so-called "rival" companies are just like them, trying to serve their community. Company leaders are identified by each group and rotated, allowing each person a chance to serve in a leadership capacity. The social bonds established in training environments that target students' social needs last for years, long after the student completes the program.

Esteem needs can also be described as ego needs. Instructors must take care to ensure that all students are made to feel important and that each is a valuable part of the training environment. Each student will have areas in which they excel and other areas where they may need improvement. Instructors must take care not to humiliate students in a way that affects their self-esteem in a negative manner. Remember to praise in public, and correct or reprimand in private. It only takes a moment to damage someone's self-esteem, and nearly a lifetime to repair it.

Mutual respect must be exercised and enforced among both students and instructors. Ensure instructors maintain a positive attitude at all times. Some instructors have their own esteem issues and bring negativity to the training program. These instructors must be screened out and separated for further instructor training before having a significant impact on the students. The training officer has the responsibility to ensure only the best instructors make it in front of a classroom. Monitoring instructor and student feedback will help identify instructors who present a less than positive attitude.

Self-actualization needs refer to the realization of success. When a student enters the volunteer fire department, they have visions of heroic firefights and rescues. Successful training programs provide the knowledge, skills, and abilities to realize this dream. The Army's slogan, "Be all that you can be," illustrates well the theory of self-actualization. Training successes must be celebrated in a way that builds pride in accomplishment and causes the student to seek out that feeling of success time and time again.

Students find pride in completing programs known for high standards and principles. Maintaining such a program will help the training officer perpetuate motivation in the student in the future. The more volunteer departments lower standards to appease their volunteers, the more they devalue their program and the pride associated with completing it. Training programs that allow substandard students to squeak by will only draw more of the same. Soon the program will develop a reputation for accepting those who would not make the grade elsewhere and become a safe haven for mediocrity and slothfulness. Programs that push the envelope of performance and weed out those who fail to perform will attract the best volunteers and students. Quality volunteers will seek out the challenge, and those who were transitional in their commitment to excellence will be influenced by quality students, rather than being drug down by the others. This may seem an unlikely method for success in today's volunteer system, but it is one that has proven successful time after time.

Capitalizing on self-actualization can also be accomplished by rewarding the student at the end of their training. Many students complete their basic firefighter certifications with little or no fanfare, yet completing hundreds of hours of training is a great accomplishment. Earning the title of firefighter must be held in the highest regard and the entire department should celebrate this rite of passage. Formal graduations, a ritual normally reserved for career departments, are excellent ways to celebrate these achievements. Graduation ceremonies allow the student an opportunity for their families, friends, and peers to honor their accomplishments in a way that facilitates pride. In fact, many volunteer departments can create graduation programs that would make most career departments jealous, as they have more freedom with their resources and can quickly gain community support.

Other awards can be created to further enhance a student's pride in their accomplishments. The Top Graduating Student and Most Improved Recruit are simple awards that offer ways of recognizing achievements. This theory of motivation can even be extended beyond the training program to graduates in the field.

Creating awards such as the Honorable Duty Award, given to those exhibiting the high standards and morals that are part of the training process, expand the reach of the training program, even beyond the classroom.

Understanding the underlying theories behind student needs provides the training officer the tools they need to enhance the student's chances of success in the modern fire training program. Most programs do well with the most basic needs, yet few go all the way to incorporate social, esteem, and self-actualization needs. The training officer should ensure their training program takes a holistic approach to the student motivation to ensure they get the most out of their program.

Herzberg's theory of dissatisfiers

Frederick Herzberg applied his studies toward the identification of motivators and dissatisfiers of people on the job. In the 1960s, he asked people to describe specific aspects of their jobs that made them feel satisfied or dissatisfied. Herzberg's "two-factor theory" separated responses into two categories: hygiene factors, meaning factors that influence levels of dissatisfaction, and motivators, or factors that influence levels of satisfaction. Figure 2–4 contrasts Herzberg's traditional list of common dissatisfiers against typical dissatisfiers in the modern volunteer fire service.

Hygiene Factors (Dissatisfiers)	
Herzberg's Dissatisfiers	**Fire Service Dissatisfiers**
Working conditions	Time demands
	▪ Increased call loads
	▪ Increased training requirements
Pay and security	Volunteer departments
	▪ Little or no compensation
	▪ Increased use of combination
	departments
Poor policies	Poor policies
Interpersonal relationships	Personnel conflicts
	Leadership problems

Fig. 2–4 Herzberg's Dissatisfiers in the Fire Service

Hygiene factors refer to working conditions, pay and security, company policies, supervisors, and interpersonal relationships. When these factors are applied to the volunteer fire service, they outline the issues that drive down volunteer retention and hamper recruitment efforts. Inappropriate or unfair polices, poor supervision, and failure to provide a sense of worth and safety are common responses to why volunteers stop volunteering.

During the 2000 National Volunteer Fire Summit, held at the National Fire Academy in Emittsburg, Maryland, reasons why people do not volunteer were examined to help identify some of the key concerns of our current volunteers. An examination of some of these reasons helps to define the hygiene factors specific to the volunteer fire service.[2]

Time demands

The emergency call volume of some volunteer stations has become staggering. In the past, volunteers viewed duty at the fire station as social time, allowing them to converse with friends and have some good old-fashioned fun. For many volunteer fire departments, those days are gone. It is not uncommon for volunteer fire stations to run thousands of calls annually, as opposed to the couple hundred of the past. The typical volunteer firefighter (male, in his 30s, with a wife and kids) could once look forward to duty night at the fire station as a way to get some quality sleep without the usual interruptions that come with family life. Today, that same firefighter could easily run three or four calls throughout the night, and then go to his full-time job the next day. Instead of the "get-away" that the volunteer fire service once was, it now adds pressure to families who are already running on full throttle.

With increased call volumes come other requirements, such as committee assignments and training requirements, etc. Like most other facets of life, the pace only seems to increase with each passing month. Parents often must take jobs, and in some cases multiple jobs, to maintain the lifestyle to which they are accustomed. Any time allocated outside the family adds stress and pressure in ways that cause a choice to be made between public service and the family unit. To ensure the emotional and mental health of the firefighter, the volunteer fire department should promote family before volunteer service. At no time should volunteers be ridiculed or teased for taking time off from the firehouse to be with their family. Departments who steadily continue to ask more from their volunteers without addressing the impact on their personal lives drive a wedge between the family and the department. Without at least acknowledging the commitment the volunteer makes, time demands quickly become a hygiene factor, a reason for dissatisfaction with the volunteer fire service.

Personnel conflicts

Coming into to the volunteer fire service can be an intimidating experience. Volunteer departments can be *clique*-oriented and new members may feel that they have to prove themselves to be accepted. Those who do not easily adapt to such situations may quickly be dissatisfied and serve only a short period. The training officer and department leaders must ensure that new members are comfortable and have a fair chance to blend into the department. Many volunteer departments have wondered why they are seeing declining recruitment numbers, yet they immediately start in on new members with nicknames and typical firehouse humor. Though a fire service tradition, this type of humor must be measured to ensure that volunteer members are not their own worst recruitment enemy.

Other conflicts may arise among department veterans that can quickly cause dissatisfaction. Most volunteer veterans can name cases in which personnel conflicts caused volunteers to quit because of false statements or rumors. Troublemakers usually have an agenda different than that of the leadership, and unfortunately, must be removed from the group. If a few people are causing a negative impact in the department and creating a feeling of dissatisfaction among the volunteers, for the benefit of the group, they should be asked to leave. This is a bold statement for some departments, but it is important to realize that not everyone is a good volunteer. Only the best should be accepted into the department and those who do not meet departmental standards only bring down the rest of the group.

Leadership problems

Many of today's potential volunteers work hard during their day jobs and are quick to notice ineffective leadership or management. With all of the troubles and politics in their full-time jobs, why sign up for more abuse at the hands of an in-experienced or overly complacent chief? Poor leadership is the cancer that is slowly driving the volunteer fire service into extinction. Many recruitment drives bring in new members only to have them discouraged by ineffective leaders who can't seem to get the job done. The volunteer fire service claims to be in desperate need of new volunteers, yet we can't seem to get the paperwork done for potential members in a timely fashion. This sends the message to our new recruits that they are unimportant, and we really don't care if they volunteer or not. They can get this kind of treatment at work; certainly they will not keep coming back for more abuse for free!

It is vital that our leaders be trained in basic leadership and management styles if we expect our volunteer service to survive in the future. As the volunteer training officer begins to assess the needs of their department, they may find that they must first focus on the current members of their department, rather than targeting recruit members as the priority. It is impossible to train quality recruit firefighters when your volunteer management is constantly driving them away. Training your leaders will be difficult and will take a great deal of time. The training officer will be challenged to find ways to coax them into training if the officer expects them to attend. They are the chiefs; surely they already know everything there is to know. The instructor must train them in spite of their attitude or experience. Management and leadership training is a constant maintenance that must be performed for your department to ensure that everyone stays focused on the mission.

Poor polices and procedures

A department whose operating rules do not facilitate fairness and understanding among its volunteers finds itself with a high sense of dissatisfaction. Verifying that grievance procedures are in place helps to ensure fairness. Volunteers, who feel that they have not been treated fairly, quickly become dissatisfied and tend to have short careers in the fire service. Don't be afraid to put the bylaws back on the drawing board if they are not volunteer-friendly. The focus should be on family, then community service, and they should provide grievance procedures for volunteers when necessary.

Increased use of combination departments

With all of the growth and demands on the modern volunteer system, some departments find themselves hiring at least some paid firefighters to supplement their resources. Balancing these relationships between the paid and volunteer sectors is sometimes a powder keg that overshadows the reason both sides were there in the first place and can quickly cause dissatisfaction within the department.

Professional standards are the key to balancing the relationship between paid and volunteer firefighters. Volunteer departments considering supplementing their services with paid personnel must establish a consistent training level for firefighters in both paid and volunteer categories. Animosities develop between the two groups when there are two separate sets of rules. Double standards can lead to dissatisfaction and may cause a decline in volunteerism if not addressed.

Herzberg's theory of motivators

Herzberg's motivators refer to personal growth, the work itself, responsibility, recognition, and achievement (Fig. 2–5). This concept describes all things that are good about the fire service. The ability to be a part of something important and to have an opportunity to advance and be recognized are at the root of the satisfaction felt by our volunteers. Such things promote long careers in the volunteer service, rather than the average two- to five-year stint.

Motivating Factors	
Herzberg's Motivators	**Fire Service Motivators**
Personal growth	Pathway program ■ Job descriptions ■ Fair process
Work itself	Focus on the mission
Responsibility	Delegation of authority
Recognition and achievement	Awards and recognition ■ Daily recognition ■ Special recognition

Fig. 2–5 Herzberg's Fire Service Motivators

Departments that capitalize on the motivators tend to be the more progressive and successful organizations. Allowing volunteers to focus on the job of helping others, shielded from the political hoopla found in many volunteer departments, is the key to modern fire service retention. Providing clear career paths that outline opportunity, including the process and requirements for advancement, facilitates healthy relationships and helps focus efforts on the department mission.

Campbell County, Wyoming has developed a unique approach to training that focuses on motivators rather than dissatisfiers. They developed the Pathways program, designed to offer each member of the department clear choices of career paths within the volunteer system. Everyone from cadet firefighter to associate member has the opportunity for advancement, based on training and achievement. Their unique approach enhances their recruitment and retention statistics and creates an increased feeling of satisfaction among their volunteers. The key to Campbell County's success is their enhancement of personal growth for all of their volunteer members. Each position in their department has a job description that outlines the requirements, duties, and responsibilities for that position. This type of approach creates a system that is fair and obtainable by all volunteers, depending on their specific area of interest.

Most noticeably missing from Herzberg's list of motivators is money, which he actually listed as a hygiene factor. Herzberg found that money served as only a temporary satisfier, appeasing complaints in the hygiene category for only a short period. It was found that raises failed to satisfy employees over the long term if they remained dissatisfied in the hygiene category. This theory can be applied to effectiveness of some incentive programs. Incentives will fail to improve retention efforts if volunteers are unhappy with the hygiene factors associated with the department. When conducting the training needs assessment, it may be necessary to address these hygiene factors through training to save the department from self-destruction. When recruiting new volunteers, quality recruits will quickly recognize these inefficiencies and will be less likely to volunteer in these conditions. Combining decreases in recruitment with a highly dissatisfied volunteer force is a sure way to expedite the need for paid personnel and possibly the end of the volunteer fire department.

Both Maslow's and Herzberg's theories should be considered when accessing learner needs. Each theory of needs can be applied to the fire service in ways that will enhance the training officer's ability to provide training for the department. Though each need addressed can seem small when singularly examined, together, the needs, motivators, and dissatisfiers have a great impact on the volunteer training system. Each will help guide the training officer toward creating a progressive, quality training program. When faced with a difficult decision or you just can't seem to come up with a good answer, reviewing these theories can offer guidance that will lead to quality and performance. Understanding volunteer needs, wants, and motivation is the key to providing training that meets the individual needs of the volunteer.

Job Needs

Once the training officer develops an understanding of individual needs and how they impact the training system, the officer must then assess what needs are required for the job. Job needs are based on statistics and fact, rather than emotion and want. They are intended to reflect the work performed by the department and the working conditions faced by firefighters. Such training needs are identified by conducting a job analysis on the various levels of the department, a task that will require some time and effort. Volunteer training officers may choose to conduct a formal and complex job analysis or an informal assessment based on simple interviews and observations. This choice depends on the size of the department and the time and resources available.

Formal job analysis

A formal job analysis of a volunteer department consists of assessing the duties and tasks the firefighters perform on a regular basis, the type of environment, tools and equipment, relationships, and requirements that impact the job.[3] By evaluating each component of the job analysis through aggressive data analysis and process management, the training officer can develop a comprehensive assessment for addressing the department's training needs. To conduct such an assessment, there are some basic areas discussed in the following section that the training officer should address.

Duties and tasks

Assessing duties and tasks of the volunteer firefighter begins with determining the knowledge, skills, and abilities, or KSAs as they are known in the human resources industry. These KSAs serve as the basis for the *firefighter job description*, a term not common in the volunteer fire service. It is often assumed that volunteer firefighters will simply fight fires, but there is far more involved than just that. Everything from EMS to technical rescue has today's volunteer firefighter doing far more than in the past. The department must create a job description outlining the scope of responsibilities and required education for the volunteer firefighter. In some cases, the required education might be determined by state or local governments, while in others, it may be established by the department itself. In either case, these requirements serve as the baseline education for both the firefighter and the training system; yet, many people know that the skills required of a firefighter go far beyond their experiences on the training ground. It is vital that the training officer fully understand the job of the firefighter to ensure that the training program meets the individual needs of the volunteer, while simultaneously meeting the needs of the department and the job.

The environment

The environment consists of the conditions in which the volunteer firefighter must perform and varies by department because each has its own set of challenges. Assess target hazards to see if there are special skills required that are lacking in the current program. High-rise training is a good example of such needs. Many departments believe that because they have no 20-story buildings in their district, they have no need for such training. Some training officers know that the tactics used in high-rise operations can be applied to certain suburban and rural settings, such as apartments or large areas. The training officer must recognize these tactical needs based on an assessment of the environment and then ensure that the required instruction becomes a part of the training system.

The environment can also include people who are hostile in the workplace. Is the environment around the station pleasant? Is it comfortable for volunteers? Training can impact these conditions by providing education in diversity and tolerance if necessary. Some training officers may view this as too "touchy-feely" for their fire department, but failing to address such issues long-term can cause reduced retention statistics and possibly civil and legal actions against the department. Also consider the people encountered by the department through service delivery, as that may present special training challenges also. For example, language barriers may present a training challenge, requiring the training officer to go outside their normal regimen of classes. At first, language training may appear to have no tactical value, but experience will show that such barriers can have a huge impact on the fire-ground operation.

Relationships

Relationships refer to the effectiveness of supervision and communication between members within the department. It is designed to measure the quality of supervision and leadership and may help to identify areas that need training. Although asking department members how they feel about their leaders can be a volatile question because not everyone will be happy, it will nevertheless help identify some areas needing improvement that can be validated. The training officer can use this survey to factor leadership and supervisory training into the overall training mission.

The transition from peer volunteer firefighter to company officer can be a difficult one. In an instant, personnel can go from "just one of the guys" to the boss. Battalion Chief Chase Sargant of Virginia Beach Fire and Rescue and Spec-Rescue, International developed a program entitled, Buddy-to-Boss, designed to provide training and education in negotiating this difficult transition. This transition has long been overlooked by the fire service and the training officer must ensure guidance is available to all personnel to meet such a looming challenge. This course is a great example of developing training to address specific areas identified through needs assessment.

Requirements

Requirements call for a review of recognized performance standards that apply to the fire department, whether they be through local code, ordinance, or professional standards established by the fire service. This review helps measure actual performance against written expectations and gives the training officer a target for future training programs. Some requirements may be law, offering departments little choice regarding compliance, while others must be adopted by the department to become official. *NFPA 1001, Standard on Qualifications for Professional Fire Fighters* lists categorically the requirements for becoming a firefighter. Even though the training officer must follow a curriculum determined by another organization, they should still review the NFPA guidelines to ensure that their training program is compliant with minimum standards. On occasion, there may be objectives or skills overlooked by the local training authority. To ensure compliance with all requirements, determine which ones apply and use them as a yardstick to determine the current performance of the department.

Gathering the information

When using a formal job analysis, the training officer must develop data collection methods to gather information. The process should include a review of statistical information, a formal written survey, and interviews to develop a sample of the current training needs. Each collection method has its protocols and requires time to complete. The training officer should allow a minimum of six months to conduct a formal job analysis to identify training needs, as some of the processes require developmental time and response periods.

A review of statistical information is conducted by reviewing any hard data available within the department. A review of training records to identify what percentage of the department has the basic level of training is a good start. Then an assessment of advanced training to identify needs and trends may identify deficiencies that should be addressed. Reviewing computer-assisted dispatch (CAD) records may also identify needs if they contain benchmarks that identify how the department performs on the fireground. Identifying on-scene and response times are common practices for most departments, but the training officer may interject even further criteria such as "water-on-the-fire" benchmarks. Using this verbal time stamp over the radio and noting it in the dispatch information, the training officer can look for trends, for example, too much time needed to deliver fire streams after arrival. Much can be learned from this data and the training officer can refer to national standards, such as NFPA 1720 to see if they are assembling adequate staffing in the appropriate amount of time. After all, much of the time spent prior to fire attack is used to assemble people and deploy hose lines. The water-on-the-fire call typically comes seconds after the crew enters for most room and contents fires.

Formal surveys can be conducted to sample the individual views of the volunteers as to how well they feel prepared for duty. The challenge in this data collection method is obtaining a broad enough sample to ensure an accurate assessment. For departments of 30 members, the training officer would need to provide each member a survey and would need a near 100% return to effectively assess the group. Even with a 50% return, one could only hope for 15 completed forms, whereas, departments of 500 might get back as many as 250 responses. Though the sample percentage of the smaller department is equal to that of the larger department, the smaller department's results may be inaccurate as it is possible to find a political clique, or group, in 15 people; whereas, this is less likely with more than 200 responses.

For larger departments with a broad membership base, a survey can be created and deployed effectively. A scoring method must be established to convert the individual responses to hard data. Conversion works best using yes/no questions or ranking items on a scale because this system facilitates quick scoring and converts to data easily. It is important, however, to include a comments section on the survey to enable the volunteer to speak freely. The survey can provide data such as the following: "Sixty percent of the department feels they are unprepared to respond to the emergencies common in their department." This information would send up a red flag to the training officer and prompt them to look for deficiencies in the basic level of training. In the cd-rom included with this text, you will find a sample Needs Assessment Questionnaire.

Once the survey is created and deployed, the training officer must allow time for response. During this period, interviews of key personnel can be conducted. This group should be diverse, including volunteer recruits, firefighters, and company officers. Questions should be determined ahead of time and should be consistent for each interview. The intent of conducting the interview in conjunction with the survey is to provide the members a chance to speak their minds in person. Some members would prefer to voice their opinions anonymously through the survey, while others would rather say it verbally. The personal interview enables the training officer to assess body language simultaneously with verbal responses to develop their own feel for the training status. By using statistical information and surveys and interviews in concert with one another, the training officer develops a picture of training needs that is comprehensive and less likely to be skewed by opinion.

Informal job analysis

Smaller volunteer departments may wish to rely solely on the statistical data and interviews to identify their job needs. This works well, as it provides the same information as the process for larger departments. The training officer for a large department can only realistically interview a small sample of the department. The training officer for a small department, on the other hand, can interview everyone. Compiling this interview information with other collected data gives the training officer an appropriate picture of their training needs.

Compiling the Information

Based on the chosen method of needs assessment, the training officer can create a written report that outlines the methods used and the results from the study. The information is passed on to the department leadership for review and approval. The report can help the training officer build a stable platform from which to implement changes in the department. Many may argue that they feel the training program should be different, but it is harder to argue against facts. It is much easier for the training officer to debate a point when there is statistical data backing him up, especially when presenting ideas up the chain of command. Having conducted and documented a needs assessment of the department will give the training officer confidence and the ability to revitalize their training program and their department in an effort to ensure the survival of the volunteer fire service in the future.

Chapter Review Questions

1. What is the first step in performing a needs assessment?

2. What are three areas to consider when evaluating a department's needs?

3. List the needs described by Abraham Maslow in his "Hierarchy of Needs."

4. Describe where money falls into Herzberg's "two-factor theory" and why.

5. Why should smaller departments consider not using a formal survey to assess needs?

Answers

1. Conferring with the former training officer to establish any current training initiatives, goals, or objectives.

2. Organizational needs, learner needs, and job needs.

3. Physiological needs, safety needs, social needs, esteem needs, and self-actualization needs.

4. Money is classified as a hygiene factor or "dissatisfier" because it will fail to motivate employees long-term if other hygiene factors are not met.

5. Smaller departments will need a near 100% return on their surveys, an unrealistic expectation in data collection.

Notes

[1] Mescon, Michael H., Courtland L. Bovée, and John V. Thill, 1999, *Business Today, 9th ed.* New Jersey: Prentice Hall Publishing, 260.

[2] National Fire Academy, *Report on the 2000 National Volunteer Fire Summit,* Washington D.C.: National Volunteer Fire Council.

[3] HR-guide.com, *Job Analysis: Overview,* www.hr-guide.com/data/G000.htm, 2001.

3

A Systems Approach to Training

Once the training officer has determined the basic training needs of the department, they must next search for ways to meet these needs with the limited resources on hand. Often, the training officer may feel overwhelmed and discouraged after the initial assessment of their department's needs. This feeling is natural, especially with the growing demands placed on the modern volunteer fire department. Addressing the laundry list of responsibilities placed on the volunteer training officer requires a different approach from that of the past. No longer can the training officer shoulder the burden of training their department alone. The modern training officer must develop a training system that addresses the variety of training needs, while rejuvenating itself by creating new instructors to meet the training challenges of tomorrow. Without the systems approach to training firefighters, the training officer is left on their own, outnumbered and out-tasked.

The following five training elements make up a typical training system: (1) the instructor staff, (2) recruit volunteer training, (3) basic firefighter training, (4) in-service training, and (5) advanced training. Each phase is designed to feed the next in a way that perpetuates the volunteer system.

The Instructor Staff

The first step to building a training system is to establish a qualified instructor base. Even though this base may consist only of a few instructors in smaller departments, it is the vital link to creating a training system. The training officer should be able to find the help they need right there in their own department. Most departments have at least one or two members who have completed instructor training. Many complete the training and wonder what they can do with it because the training system fails to recognize the new instructor as a valuable resource. Only when departments are created from scratch should the training officer have to look outside the department to find instructor help. When departments are first conceived, the training officer may have to get commitments from instructors in the general area until the system has time to cycle out some new instructors. Even after that, the outside instructors may be needed to provide mentorship to the new batch of instructors.

Along with identifying instructors to assist in the training mission, the volunteer training officer must understand that they can't go it alone. Even when training a small company of 20 to 30 firefighters, the training officer will need help to accomplish the mission. The successful training officer will realize that the job is more about managing instructors and programs than actually standing in front of a classroom teaching. This realization is difficult for some, but is necessary to address a potentially overwhelming training burden.

Consider the instructor/student ratio when evaluating the training workload. Divide the members of the department by the number of instructors assigned to training. This figure can give you an idea of the demands placed on the training staff, whether there be many or just one. Actually applying a figure to the workload will help the training officer make their case for more help in future requests for resources by translating the work into a value that the bean counters can understand.

The Hanover Fire Academy found it was significantly overwhelmed by the training needs of their department. Being a department of 600 volunteers, three instructors were hired to train the masses. With a ratio of one instructor per 200 volunteers, the system was overwhelmed, making it difficult to deliver even the most basic training. The academy solved this problem by organizing their assigned instructors into coordinator roles, managing volunteer instructors throughout the department to meet the training needs. By simply changing the focus of their job descriptions and mobilizing available resources, this team brought its instructor/student ratio down from 1:200 to 1:30.

Of the more than 600 volunteer firefighters in the county, many were qualified as instructors, yet few took part in the countywide training initiatives. Twenty volunteer instructors were recruited to assist in the training effort under the guidance of the three full-time instructors. These extra instructors were recruited by explaining the facts to the membership at large. The department had seen a steady decline in volunteer activity due to the usual garden variety of reasons associated with declines in volunteer retention. Playing the statistics out over a five-year period, the training division was able to show that the volunteer department could actually become extinct if it didn't assume responsibility for the current situation and work to train and rejuvenate the department. A series of presentations made across the county helped mobilize this group of instructors to stand up and do something to help.

Another common problem among volunteer instructors is burnout. The instructors in Hanover had been mobilized several times before with only moderate success. Volunteers helped in the instruction for a short time, but disappeared quickly, leaving the training system understaffed and unable to meet the growing training needs. The training staff had to search for ways to mobilize the available instructors in the area, while minimizing burnout. The staff took a unique and somewhat unorthodox approach to motivating and retaining their instructors.

A key first step to increasing instructor retention and reducing burnout is to simply create an identity that the instructors can find pride and honor in serving. Charles Brush, in his article "How to Market Like the Marines," describes how the U.S. Marine Corps turned low recruiting statistics and high dropout rates into The Few, The Proud, The Marines, a slogan that proved to be highly successful, giving the Marines a sharp rebound in recruitment and retention.[1] The marketing approach used by the Marines is almost humorous. Their recruitment numbers were down, so they were the few; the few they had were very proud to be Marines. They turned a negative situation positive in a way that helped them achieve their goals, rather than focusing on the negative and hindering progress. This unique approach built for the Marines a reputation among its target audience of being an elite organization where only the best make the grade. This technique plays on a strategy that has worked time and again in the volunteer fire service as well: offer up a challenge and those willing to serve will find a way to meet it.

This same principle can be applied to a group of instructors by establishing an identity for the group. Adopting names like "Small Town Fire Academy" or "Anytown Training Center" will help the instructors feel as if they are part of something elite. A dedicated training site is not needed to create such an identity. Some academies operate out of a variety of locations, moving classes from station to station. Volunteer academies are not properties, structures, or fences; they are built on pride, honor, and commitment. The focus should be on establishing and marketing the achievements and abilities of the instructor team, rather than the location itself.

Some departments may have funding to provide special academy tee shirts or uniforms for instructors, while in other departments, instructors pitch in their own money to buy shirts or uniforms. In either case, the instructor team should wear their academy uniforms or shirts with pride, as they are truly an elite group. This uniform should be reserved only for those who participate and meet the standards set by the group. Allowing others to wear the uniform will dilute the pride associated with the group and must be avoided whenever possible (Fig. 3–1).

Fig. 3–1 The Hanover Fire Academy Created its Own Identity to Help Mobilize Instructors

An example of this phenomenon is demonstrated by the Fraternal Order of Leatherheads Society, otherwise known as the FOOLS, who represent the brotherhood of the fire service. The organization is reserved only for those who truly love the fire service, providing fellowship, training, and community service in chapters throughout the world. The society was started by three firefighters in Florida who believed in the proud traditions of the fire service and created an identity to further the cause. Today, that group of three brothers has expanded across the globe, as brothers and sisters who share these beliefs hurry to join the organization. By giving their ideals and beliefs an identity, they were able to mobilize a nation and a world of firefighters to stand with them in their cause. For more information about the FOOLS, go to www.foolsinternational.com. Adopting this philosophy and giving the training program an identity will do much to improve recruitment and retention of instructors. Training systems using this approach now enjoy a waiting list of instructors wanting to become part of the elite group.

Instructor development

Along with giving the instructors an identity to stand behind and be proud of, a key to perpetuating the volunteer training program in the future is providing development and advancement for instructors. With the growing demands on today's volunteers, quality volunteer instructors are even more rare than volunteer firefighters. Training officers must nurture their staffs, providing them areas for advancement and self-improvement. This can be a great challenge with funding for training already spread to capacity in normal training operations. The training officer will have to do much of the instructor training in-house, or find outside experts who are willing to teach for a dinner or baseball cap.

When developing an instructor development program, the volunteer training officer must ensure that instructor training is organized and progresses from the new to the experienced instructor. Many view completion of the NFPA equivalent for fire service instructor as adequate, and it is true that it can serve as a baseline education for all instructors. What most fail to do however is go beyond that initial training and offer additional instruction that expands on the content of the basic level instructor training. Without such follow-up training, the new instructor can be left unsure of direction and wondering how to be an asset to the department. Creating a career path, supported by advanced training education, the new instructor can be brought into the elite fold of the training organization (Fig. 3–2). The instructor development program contains basic training courses for new instructors and allows them to progress through the ranks based on performance and education.

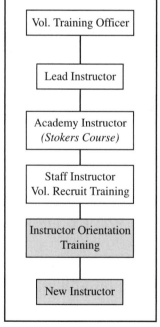

Fig. 3–2 Sample Volunteer Career Path in a Volunteer Training System

Instructor orientation

As volunteers obtain instructor status, the volunteer training officer should be prepared to describe the training program to them in detail. An orientation program should be conducted as needed to ensure that new instructors get off on the right foot. A detailed review and explanation of the training system should be conducted and procedures should be explained. The orientation should begin with introductions of the training staff and a description of the facilities. If a formal training site is available, procedures for locking and unlocking the facilities should be discussed along with cleanup assignments. A description of the various training sites should be provided and access procedures for each building identified. Instructors should be made aware of the availability of audio-visual support and access to the appropriate curriculums. Procedures for instructor assignments and emergency cancellations should be identified to ensure that each instructor knows what to do for inclement weather, major fire emergencies, and personal emergencies.

Once all of the details have been discussed, the instructors should be assigned their academy uniforms, if available, and any keys or door codes that may be required. It is important to instill a sense of pride in being part of the academy program. New instructors should feel as if they have joined an elite team of professionals focused on meeting their identified mission. Reinforce the idea that being issued these keys, codes, and equipment is a privilege and should be held in the highest regard.

Upon completing the orientation program, the new instructors are ready to begin some limited instruction under close supervision. Departments that use some method of new volunteer orientation can break in their new instructors at this level. For departments that lack some form of orientation training for new volunteers, a sample program is provided in chapter 5. Using the new instructors at this level of training enables them to ease into the front of the classroom in front of small groups, rather than before classes of 30 or more. The new instructors can teach at this level until they are comfortable and the training officer feels they are ready to move to the next level of training and responsibility.

Stoker course

Aside from speaking in front of a group, one of the more difficult skills to master for new instructors is conducting live fire evolutions. With the inherent safety risks associated with live fire training, the training officer must ensure that all personnel working on the interior of the building are aware of all safety regulations, emergency procedures, and burn guidelines. Often, new instructors become overzealous in building fires and hamper the student's ability to learn in a realistic training environment. It is important that new instructors serve an internship with an experienced instructor until burn procedures can be mastered.

The first step in the stoker course is to identify the need for such training. Repeatedly firefighters are injured or killed during training exercises due to lack of planning and experience. Review the statistics that demonstrate this risk to ensure that instructors understand the full weight of their responsibility. It should be made clear that a zero tolerance policy is in effect regarding live fire procedures. Explain further that instructors who fail to follow the guidelines in the course will no longer be allowed to participate in live fire operations.

Next, an itemized overview of the requirements contained in the *NFPA 1403, Standard on Live Fire Training Evolutions* should be conducted. This should cover the points listed for both burn buildings and acquired structures to ensure that the instructors are aware of each requirement. Be sure to point out the water supply and command structure requirements in detail. The department's emergency evacuation procedures should be reviewed and available radio frequencies identified. A complete overview of pre-burn, during the burn, and post-burn procedures rounds out the program.

A written exam should follow the lecture to document each instructor's under-standing of the burn program. The test should consist of questions regarding the *NFPA 1403* and any local procedures that may apply. The test, along with a copy of the roll sheet, should go into the instructor's file to serve as documentation in the event of a future injury or accident to protect both the training officer and the department from possible liability. Though these records may not offer protection from all liability, they will prove that the instructor was trained to operate within the guidelines of the NFPA professional standard.

A practical hands-on session should be conducted to show how the jurisdiction operates a live burn. Building preparation should be discussed, allowing the instructor to see how the building is stocked with fuel and how instructors move about the building safely. All safety features, including vent openings, window and door operations, and safe havens should be identified. The water supply plan should be discussed and set up by the new instructors to ensure that they have a full understanding of the system. Once all preparatory work is done, the instructors are shown how to conduct a student walk-through of the structure as required by NFPA standards.

Setting fires in burn buildings requires experience and finesse to master. Most burn buildings operate between 500° and 1,000°F, leaving little room for error. Instructors must realize that their job is to create fire conditions found on the job as realistically as possible. Unfortunately, doing so in a building made of concrete and steel can be quite difficult. Most houses have combustible structural components along with furniture, appliances, and personal belongings. Creating a realistic fire condition means the instructor must have a theatrical understanding of the students' perception.

Heavy smoke conditions tend to provide a more realistic evolution. As inexperienced instructors crank up the heat, they start to burn off the smoke, clearing the room to near perfect visibility. This condition gives the student a false impression of real firefighting conditions. The better approach is to concentrate on building a fire that produces heavy smoke and can be accelerated quickly as the student approaches the burn area.

Loading the burn pan should be in accordance with NFPA standards and must meet any guidelines set by the department. Most jurisdictions use pallets and straw to build the fires. This works well when done in a large pan or barrel to contain the debris. Work one-on-one with each instructor to ensure that each knows how to build a fire using minimum fuel loading while allowing air to draw through the pallets (Fig. 3–3). The straw has two functions: (1) to ignite the wood in the pallets, which is usually identified through telltale crackling of the wood, and (2) to provide a rapid flash as the students approach the burn room.

Fig. 3–3 Burn Pans Consisting of Three Pallets and a Bail of Straw Should be the Maximum Fuel Load

Once all new instructors are comfortable, fires may actually be ignited to give the new staff a chance to observe the burn operations from the *other side*. Instructors must know how to create realistic fire conditions that include dense smoke, and then accelerate the fire just as the students arrive and get out of their way all at once. This requires an ability to move rapidly and safely in a heavy smoke condition. Experienced instructors act as students and move in for the fire attack, allowing the new instructors to hone their skills under the supervision of the training officer. This type of evolution enables the training officer to observe the new instructor team to ensure that they understand proper burn operations. Assigning the new instructors to work under the direction of a seasoned instructor will help bring the new skills together and ensure a safe live fire training operation.

Typically, upon completing the stoker course, the new instructors have enough experience to feel confident enough to move to the front of the large classrooms as lead instructors. With this designation comes the responsibility of controlling the class and commanding the training operation. Most new instructors gain at least a year's worth of experience before stepping up to this level, but there will be those who excel rapidly and the training officer must recognize this talent and move those individuals forward as they are ready. Once at the lead instructor level, they can assume responsibility for complete courses and are highly-valued assets to the training officer. The training officer must take care to nurture these new instructors and see that they realize this success.

Peer counseling and mentoring

Instructors can always find ways to improve their delivery and technique, and new instructors value guidance and advice. The training officer can facilitate such growth and mentorship by implementing a few simple programs. A simple tool to enhance instructional delivery is to videotape the class. If resources permit, the training officer should provide a video camera in the classroom for instructors to use as they wish. The new instructors may then watch their presentations in private and self-evaluate their performance. Often, the instructors will be harder on themselves than others will.

Instructor Development Teams (IDTs) provide instructors with peer counseling and ensure high levels of performance. IDTs consist of three instructors of equal or higher certification who critique a new instructor's performance. Each of the three evaluators complete an instructor evaluation form (located in the included cd-rom) and turns it in to the training officer for review. The training officer then meets with the instructor privately to review the evaluations. The instructor/evaluators do not meet directly with the instructor to reduce the potential for argument and debate. Instructors appreciate positive suggestions on how to improve their delivery, and such evaluations allow them to constantly improve. This process also consistently raises the bar of performance and helps prevent the stagnant complacency that often plagues training programs over time. It is important to ensure that these evaluations are done in the spirit of quality assurance and the training officer must be on the lookout for those who have alternative agendas.

As new instructors are brought into the academy system, they should be assigned to assist more seasoned instructors. Working one-on-one with a veteran instructor allows the new instructor the time and guidance to find their teaching style and to perfect their delivery. Like basic firefighter training programs, instructor training courses are designed to give the new instructor the basics for delivering firefighter educational programs. New firefighters must work under the close supervision of a seasoned company officer, just as new instructors must work closely with veteran instructors to perfect their delivery. Assigning veteran instructors new instructors to mentor enhances the performance of new team members and provides a unique sense of purpose for experienced instructors, which improves instructor retention. Senior instructors are encouraged to pass on their experiences and pass the work ethic on to the next generation. Combining this approach with detailed training and a career path can maintain and rejuvenate instructors as needed to ensure that the training mission is a success.

Recruit Training

Volunteer recruit training is an often-overlooked process that places stress on the new volunteer. Many departments find themselves torn between requiring firefighter training before allowing new volunteers to ride the fire apparatus and trying to retain volunteers by getting them into the action quickly before they become bored and disinterested. The answer to this dilemma lies in the balance between safety and excitement. Volunteer recruits cannot be allowed into harm's way without appropriate training, but neither do they want to wait up to a year to complete the necessary certification courses. The key to addressing this issue is developing an interim training program designed to give recruits the skills they need to be safe on the fireground as well as allowing them to feel like a valuable part of the team.

NFPA 1403, Standard on Live Fire Training Evolutions can serve as a roadmap for developing such training. This standard calls for specific training prior to students participating in live fire training. The required topics are fire behavior, safety, personal protective equipment (PPE), self-contained breathing apparatus (SCBA), fire hose, appliances and streams, forcible entry, ventilation, ladders, and rescue.[2] A compact course that targets these topics offers training for the entry level volunteer that will provide for their safety and not be overwhelming to the training officer or volunteer schedule. This training serves as a bridge for the new volunteer from joining the department to finishing their firefighter certifications.

The first step is developing course objectives for each required topic listed in *NFPA 1403*. The standard calls for the training to be at the Firefighter I level or equivalent. This presents a question the training officer must answer early on in the process. Will the orientation training count toward Firefighter I certification when recruits go to the next level? Each training officer must evaluate this question, but the recommended answer is no. Depending on the delivery system, there can be up to a six-month break between initial training and Firefighter I training and the orientation training barely skims the Firefighter I requirements outlined in the *NFPA 1001*. Firefighter I and II training go into much greater detail, far beyond the requirements of the minimum standard.

Next, a course outline must be developed and audio-visual aids created. Each subject is designed to be taught rapid-fire with quick lectures on the topic, moving on to hands-on training as quickly as possible. Determine a logical sequence of subjects and be sure to include any department-specific training that may be required on the first night of training. Department orientation or infectious disease training are common first night topics.

Diversifying the orientation

Departments that have both fire and EMS transport services may wish to create career paths for volunteers based on their interests. There may also be those volunteers who wish to help, but are not interested in riding a fire engine or ambulance. Historically, this type of volunteer is lost because they don't fit well into either category. Unfortunately, this lost volunteer contingent could be a huge asset to a department as these may be accountants, engineers, financial planners, marketing specialists, etc. The orientation process can be used to capture this group and help them find meaningful duties in the volunteer department without working as emergency responders. Virtually all volunteer departments are run similarly. Funding must be obtained and managed, volunteers must be recruited, and the service must be marketed. This potential group of volunteers could be a major asset to the department president or treasurer. The volunteer system must simply find a way to capture this group and tap into their services.

The Pathways program developed by Campbell County, Wyoming, mentioned in the previous chapter, serves as a model program for capturing all facets of the volunteer service. Pathways was designed to provide career paths for every aspect of the volunteer service. It provides standard volunteer fire and EMS training, but then goes further to provide administrative and business training for the administrative side of the department and even mechanical training for those interested in vehicle maintenance. This same approach can be built into the orientation program with some basic planning.

The orientation program is designed with three basic paths that allow the volunteer to choose an area of service. Volunteers may choose from the fire path, EMS path, and administrative path. Some may choose to train for multiple paths, depending on their interests (Fig. 3–4). Each new member attends the first night of training for an orientation to the department, to learn about the department's history, and to hear about their options for volunteer service. They then select a career path for their initial training.

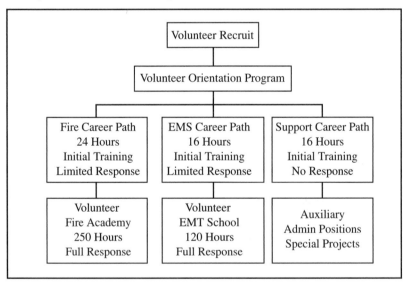

Fig. 3–4 Volunteers Should Have Training Options Based on Services Provided by the Department (EMT—emergency medical technician)

The course schedule for each path is different. The fire path consists of 24 hours spread out over a two-week period. The EMS path is 16 hours spread out over four consecutive nights, or two nights a week for two weeks. The training officer must decide which scheduling method will best suit their needs. The administrative path can be as long as necessary, depending on what opportunities the department wishes to extend to the volunteer group. Figure 3–5 offers an overview of the scheduling matrix and course outline.

Fire Path	EMS Path	Administrative Path
Orientation	Orientation	Orientation
Infectious disease control	Infectious disease control	Infectious disease control
Safety	Safety	Safety
Fire behavior	Ambulance operations	Volunteer opportunities in
PPE and safety	CPR	emergency services
Hose and appliances		
Forcible entry and ventilation		
Ladders and rescue		

Fig. 3–5 All Students Train Together in the Early Subjects and then Choose Their Specialty

Fire topics are grouped together based on what is logical and makes good logistical sense. Students receive a brief lecture on each topic and quickly move to the hands-on training. Focus is placed on what the volunteer recruit will be allowed to do on the emergency scene, for example, footing ladders, tending hose at the door, assisting with exterior ventilation, etc.

EMS topics may consist of what the training officer deems appropriate; departments that do not transport may wish to delete this section altogether. Ambulance operations provide basic safety and procedures training for working around ambulances that includes the location of key equipment and which pieces are carried in on various types of calls. Students are taught how to operate and lift the stretchers used on the ambulance, how to operate radios and emergency equipment, and anything else that might assist the crew. Cardiopulmonary resuscitation (CPR) training may also be offered, as the volunteer recruit could prove valuable in assisting the ambulance staff with chest compressions when staffing is low. Training officers may choose to expand on these topics, depending on their individual needs.

The administration path is designed to match volunteers with duties that correspond to their talents and abilities. The training officer should develop a session that outlines the business side of the volunteer department, listing the various administrative duties available. Responsibilities might include clerical support, financial guidance and assistance, fundraising, etc. For departments that maintain their own apparatus, volunteers might provide mechanical training or recruit qualified mechanics to manage a periodic maintenance program. An area overlooked by many volunteer departments is using volunteers with marketing backgrounds to assist with volunteer recruitment and retention. Training officers must use their imaginations to fully capture the potential of their volunteers.

Another option for specific recruit training is allowing new members to ride the fire apparatus after completing an equivalent part of the Firefighter I certification program. This approach is preferred for departments that start Firefighter I or academy programs more frequently than every six months. This option is normally reserved for large volunteer departments that have the staff to support such training.

It must be stressed that this training serves only as an orientation to the fire service and in no way replaces the Firefighter I and II courses described in *NFPA 1001*. Upon completion of this training, the volunteer recruit may only be allowed to participate in a support role for exterior operations on the fireground. These stipulations must be reinforced throughout the program and to other members of the department to ensure that no one is given a task for which they have not been suitably trained.

Recruit training programs must be scheduled in a manner that is consistent with departmental recruitment efforts. Offering the program on a six-week rotation makes the program available on a regular basis for new recruits and offers instructors a break between programs to work on other projects, or simply to rest. A 24-hour format gets the required training delivered in the most compact schedule. A Monday through Thursday night schedule over two weeks is also recommended instead of dragging the training out for months. This allows the recruit to get the training completed rapidly and lessens the blow to both the recruit's and the department's schedule.

Admittedly, having classes Monday through Thursday nights for two consecutive weeks can be demanding for some volunteers. Creating a process by which classes can be made up when missed will help the volunteer meet the demand, even with the most hectic schedules. Creating multiple instructors who are qualified to teach the program and making them available to the students will ease volunteer stress and give the program the flexibility required in a volunteer system.

Required attendance on the first and last nights of class is the only true attendance requirement. The first night allows the volunteer to obtain important orientation and safety information. The last night consists of a written and practical exam to ensure the volunteer has met the course objectives. Volunteers who cannot attend the regular schedule may then work with a qualified instructor one-on-one to complete the course objectives. Once completed, they can report on a scheduled testing night to complete the program. This plan can take one month or as many as 12 months to complete, depending on a recruit's personal schedule. In the past, the majority of volunteer recruits has completed the program as a group, finishing the two-week program to get it over with quickly, so that they could move on to other things.

Those who join the department with previous firefighter certification have the option of challenging the final exam after completing the first night of training. They simply present a copy of their certification, attend the safety and orientation training, and are issued protective clothing and sent to the station.

Using the recruit training to develop new instructors, as well as new volunteers, serves as the foundation of the training system. By implementing the recruit program, the training officer is better equipped to support the department's recruitment efforts and provide training that will help with retention efforts. Rotation schedules create a pool of new volunteers who will be ready to progress to the next level of firefighter training and remain an integral and important part of the volunteer fire department.

Basic Firefighter Training

Basic firefighter training serves as the workhorse of the fire training system. It typically is the largest of the training areas and is based on *NFPA 1001*. Luckily, there are prepackaged programs on the market that offer the training officer canned curriculums for Firefighter I and II training. Determining which program to use is often governed by the authority that issues the certification. The training officer should check to see which program is recognized for their jurisdiction. In some cases, the training officer may have the luxury of choosing the curriculum package; in other cases, the officer may lack the funding to purchase one. In either situation, simply remember to follow *NFPA 1001* to ensure that the program meets the minimum requirements.

Based on the recruitment for the department and the training needs assessment, determine how often basic level training will be required. As a rule of thumb, a program should be planned for every 30 volunteers recruited per year. This addresses only new members and does not account for volunteers already in the system who lack the basic training. Figure an additional program for every 30 volunteers in the department lacking training.

For some training officers, this figure may be staggering when establishing the number of programs that need to be delivered. It is unrealistic to think that the training officer can run multiple programs at once, as most lack the resources for such a commitment. To project how long it will take to meet the training objective, the training officer must develop a plan that offers programs as resources permit. For example, the Hanover Fire Department found it had several hundred volunteers who had not completed the training requirements mandated by policy. Rather than expelling these volunteers for noncompliance, a plan was presented that incorporated basic training for new volunteers and current members. It was proposed that Hanover run three academies per year for a period of six years to meet the final training goal. Even though the academy is still working to meet this goal, with support from the instructors, the department will eventually get back to its training agenda.

The department must identify what will be considered basic training for its firefighters. To do so, the training officer must imagine what the end product of the training program will be and work backwards from that point. The minimum recommended training for basic level firefighters is as follows:

- Firefighters I and II (*NFPA 1001*)
- Hazardous materials, operations level
- Firefighter survival training (Mayday! Firefighter Down, Saving Our Own, Get Out Alive, etc.)
- Emergency vehicle operator
- Vehicle extrication

The training officer may also wish to offer basic pump training, depending on the needs of the department.

Preparing basic training to meet the intent of the *NFPA 1001* should be projected to require a minimum of six months to complete on a volunteer-based schedule. Even at this pace, the student will be required to attend an average of two weeknights and every other Saturday to complete the program. The preferred method for addressing basic level training is the academy format, outlined in the next chapter.

In-Service Training

In-service training has long been a troublespot for the volunteer training officer. There are some resources available to help the training officer address this issue, but some stark realities must also be understood. Some training is required for the job and there isn't much that can be done to make it more exciting. Volunteer firefighters must accept that these courses are an unpleasant fact in the fire service. The training officer must do all that can be done to make it as exciting as possible, but there will be those classes in which there is but so much that can be done.

In-service training falls into two general categories: high-frequency/low-risk and low-frequency/high-risk. Each has a specific place in the fire service and must be considered in the training system.

High-frequency/low-risk

High-frequency/low-risk characterizes training topics that occur frequently on the fireground and carry less risk based on frequency of occurrence. For example, taking a blood pressure happens daily in the modern fire service and carries little risk to the firefighter when appropriate precautions are taken. Because it is an almost daily occurrence, refresher training may not be as critical, since firefighters carry out this task regularly.

Some training requirements for this type of training are driven and mandated by regulatory agencies, such as the Occupational Safety and Health Administration (OSHA), and must be carried out per their directions. Volunteers and training officers alike should recognize this training as a requirement and accept it as part of the job. Imagination will go far to spice up the recurrent personnel protective equipment (PPE) training, but after 20 years as a volunteer firefighter, there is only so much that can be exciting about PPE training.

Low-frequency/high-risk

Low-frequency/high-risk describes the core of the mission in the fire service and must be a priority for the in-service program. Situations that occur rarely, but carry a high risk to firefighters should be practiced most. In today's fire service, this describes even the basic working fire, as more time seems to be spent preparing for emergency medical calls, hazardous materials incidents, and the like. Firefighters require time to sharpen their hose handling, forcible entry, and rescue skills to ensure the safety of both the public and firefighters during fires involving structures.

An excellent source for creating in-service drills is *NFPA 1410, Standard on Training for Initial Emergency Scene Operations.* The standard serves as a guide for providing in-service training on topics such as handline performance, master streams, automatic sprinkler support, and truck company operations. The standard provides evaluation methods and even offers sample evolutions, including diagrams of how to lay the evolutions out in the field. It suggests methods of evaluation that establish standards of performance for the department that are measurable and help to create a benchmark for improvements in service delivery.

Performance data gives the department leadership ammunition when competing for funding or conducting the annual fund drive. They can present departmental services in such a way that politicians, citizens, and local businesses can understand them and find them tangible, thus greasing the wheels for additional funding and support. Without this data, it is difficult to describe what the department does in ways that translate well to the business world. Many people still think that volunteer firefighters sit in front of the station and play checkers all day. (If this is the case, certainly move the game inside the station!)

Advanced Training

Advanced training refers to topics such as officer, technical rescue, and hazardous materials training. Each may have a great impact on the volunteer training officer, depending on service delivery for the department. Training officers are tasked to incorporate these disciplines into their training programs along with the variety of other training needs, making it very difficult for the training officer to manage. These types of programs often require specialized equipment and instructors that are often expensive and may not be readily available to the volunteer department. The training officer should refer to their needs assessment to determine whether such training is actually warranted.

All departments will have the need for officer training. The training officer must evaluate the department to determine what level of leadership training is needed. Officer training is divided into two main subgroups: leadership training and management training. Each has its own characteristics and needs in the fire service. The old saying that *you manage things and you lead people* rings true for this training need. Management training refers to administrative skills needed by the fire officer, while leadership refers more to the tactical ability of the officer to lead troops on the fireground. Once again, the training officer can look to the administrative resources available in the department to find assistance with the management aspects of the program. For leadership assistance, the training officer may have to look throughout the fire service to find speakers and instructors who can meet their needs.

The term *born leaders* may ring true to some degree, as some have a knack for leading people. All leaders will have to work to hone their skills to be on top of the leadership game. Search for leaders who have made a positive impact on their department and the fire service as a whole, and get them to instruct your officers. Look for what works and capitalize on it. Effective leaders find their abilities to be like that illusive, perfect golf swing. True leaders are constantly working on their craft and cherish their responsibility by forever gathering new information and techniques. The training officer must do whatever it takes to bring this fire to their troops and inspire their fire officers.

Using the system to ensure success

By ensuring that each part of the system is working properly, the training officer can move volunteers from the recruit level, to the basic firefighter level, and then on to advanced levels such as fire officer. The systems approach will naturally create and nurture new instructors to ensure that the process is sustained into the future. A breakdown in any part of the system will eventually cause the volunteer system to break down and diminish its ability to recruit, train, and retain volunteer firefighters.

Chapter Review Questions

1. List the five basic components of the training system.

2. What well-known marketing concept described in the text can be used to revitalize the instructor pool and make training something that instructors want to do?

3. What NFPA standard serves as the outline for the stoker course?

4. What three career paths should recruit orientation training cover for departments that provide both fire and EMS?

5. What is the recommended maximum enrollment for basic firefighter training?

6. In-service training can be broken into what two classifications?

Answers

1. The five basic components of the training system are (1) the instructor staff, (2) recruit volunteer training, (3) basic firefighter training, (4) in-service training, and (5) advanced training.

2. The marketing like the Marines' concept of the few, the proud, the marines.

3. *NFPA 1403, Standard on Live Fire Training Evolutions.*

4. The fire path, EMS path, and administrative path.

5. Thirty students.

6. High-frequency/low-risk and high-risk/low-frequency training.

Notes

[1] Brush, Charles, 2000, "How to Market Like the Marines," *Fire Engineering Magazine* (January).

[2] National Fire Protection Association, 1997, *NFPA 1403, Standard on Conducting Live Fire Training Evolutions.*

[3] National Fire Protection Association, 1997, *NFPA 1410, Standard on Training for Initial Emergency Scene.*

4

The Academy, Part 1:
Laying The Training Foundation

The phrase, "But we're just volunteers…" has become a frequent copout for training requirements, creating a victim mentality among the volunteer fire service. This *poor me* attitude has created a sense of negativity and failure among the ranks and has caused training officers to go back to the drawing board to find ways to meet their mission. Pulling the volunteer fire service out of this slump will require a return to the core values and training methods of the past. This chapter will outline the first step in creating a volunteer fire academy designed to deliver basic firefighter training in a manner not usually associated with the volunteer fire service. Training officers may use as much or as little of the program as they feel will benefit their department, since going to a full academy process will take time to implement.

What some may view as a new, complex training method is simply organizing what has historically worked well in the fire service into a program that is tailored to the volunteer system. Using an *I dare you to succeed* approach to training, instructors challenge volunteers to excel in the program while promoting personal achievement and excellence. Firefighters by nature tend to enjoy a good challenge. The academy program presents the training challenges and offers a support network for success while focusing on performance. True, not everyone in the community is cut out to be a firefighter by today's standards, but they can certainly participate in a support role. For those who risk their lives fighting fires, returning to the high standards of discipline and performance of the past is the key to volunteer success in the future.

The academy format provides volunteer training officers with a systems approach that is detailed in policy and procedure while promoting high performance standards and teamwork. The term academy describes how the training system is organized rather than the more traditional geographic or physical location. As discussed earlier, fire departments do not need a formal training facility to create a fire academy; they only need recruits, determination, and organizational skills to be successful. Some of the most successful volunteer academy programs are created through the cooperation of multiple jurisdictions that share resources to provide basic firefighter training to their volunteers. There is no training center or central training ground, just a spirit of professionalism and determination.

When the need for basic training has been identified, the training officer should consult with surrounding jurisdictions to determine whether there is a similar need in those communities. More resources may be available by combining efforts rather than acting alone. This may give the training officer increased access to fire apparatus, training facilities, and technology. Along with greater access to tangible resources, consolidation may also enhance the organization's political influence because they provide more of the local service, and thus, carry more political influence.

Creating a Written Program

Whether acting as a single department or in cooperation with other jurisdictions, it is vital that a written program be established outlining the specifics of the program. The written program serves as both a detailed agreement for sharing resources among jurisdictions and the rules for both instructors and students to follow. Without a comprehensive written program, academies will fold at the first challenge of their authority. With the length of time required for completing basic firefighter training and the variety of personalities involved, a serious challenge to academy authority can be expected. These challenges can have a negative impact on department performance and can even have negative legal implications.

The ideas for some of the more successful academy programs were hatched over pizza or sitting around the firehouse by instructors who simply wanted to improve volunteer training. These instructors learned firsthand the need for a written plan that detailed the duties and responsibilities of each person involved. There is the story of a student or instructor who challenged the program for each rule or requirement in the program. The program described in this chapter is a reflection of those experiences from instructors throughout the fire service. The program is supplemented by the Aurora (CO) Fire Academy's *Recruit Academy Guidelines*, which are included as a supplement on the CD-ROM. Though these guidelines were written for career academies, their policies and procedures have been adopted for the volunteer fire service with much success.

A comprehensive written program can be organized into two main categories: (1) administrative guidelines, and (2) student requirements. Each plays a major role in ensuring the success of the academy process.

Administrative guidelines

The administrative guidelines should detail the authority of the fire academy and the relationship between the jurisdictions or companies involved. This is vital to the command structure and should be detailed in the guidelines. Some academies use a single coordinator to provide guidance and leadership while others incorporate a board of directors into their leadership. Each system has its benefits, and training officers must determine which method will best suit their needs.

Using a board of directors is beneficial when multiple jurisdictions are involved in the basic training process. A board enables each jurisdiction to have input into decisions regarding the direction and methods used by the academy. This structure can offer a broad base of experience through its multiple members who often have vast experience in the fire service. Using multiple members also helps to share the burden of overseeing programs that may last up to six months. Though the board approach works well politically, it sometimes can be sluggish in ruling on training issues, as all of its members must be assembled to make a group decision.

When using the single program coordinator, the person charged with running the academy must shoulder the burden for a long period and is responsible for the overall leadership of the program. All decisions rest with the coordinator, usually at the direction of an administrative chief, which can result in quick decisions regarding the training program. It is important that the coordinator have the appropriate background and experience to lead the training program in a manner that is consistent with fire service standards.

Since few volunteer departments are capable of offering self-certification to National Pro Board standards, most departments obtain certification through an outside agency. The administrative guidelines should specify this relationship and the process by which such certifications will be obtained. The written training program should be presented to the certifying agency to ensure that it is in full compliance with all certification requirements.

Outlining job duties associated with basic firefighter training is important in achieving efficiency and providing opportunities for advancement among instructors. Job descriptions should outline specific responsibilities and the requirements for holding particular titles. Figure 4–1 offers an example of job responsibilities associated with basic firefighter training and the requirements needed to serve in those positions. This example is for a large-scale volunteer fire academy. Training officers may find they can combine some job duties and titles, depending on the size of their department.

Position	Duties and Requirements
Program Coordinator	The captain assigned to training, responsible for the overall management of the training academy under the direction of the division chief. Responsible for instruction and program development, training budget, management of resources, personnel, and direction of fire training courses.
Assistant Coordinator	The lieutenant assigned to training, responsible for assisting in the management of the training academy under the direction of the program coordinator. Responsibilities include instructor development, management of personnel and resources, and serving as the student liaison.
Volunteer Training Officer	Responsible for course delivery of the basic firefighter training program to include instructor assignments and quality assurance. Must be a minimum of Fire Instructor III and Fire Officer II.
Lead Instructor	Responsible for delivery of specific course subjects within the training system. Ensures that programs are delivered in a manner consistent with academy standards and maintains all associated records. Coordinates training activities of other instructors to ensure practical objectives are met. Must be a minimum Fire Instructor II.
Academy Instructor	Responsible for course delivery under the direction of the lead instructor. Supervises individual companies during hands-on instruction. Must be a minimum Fire Instructor I and must be recommended by the volunteer training officer.
Staff Instructor	Responsible for assisting in the delivery of training, primarily recruit firefighter training, through the orientation program. Must be a minimum of Fire Instructor I.
Specialty Instructor	Personnel outside the fire department deemed to be subject matter experts by the program coordinator.
Stoker/Assistant	May assist on the training ground and on burn evolutions. Must be minimum of Firefighter II and must graduate from the stokers course to participate in live fire evolutions.

Fig. 4–1 Sample Job Responsibilities for the Volunteer Fire Academy

The mission and honor statement

Once the basic duties and responsibilities are identified, the academy must identify and communicate its mission. Some departments have mission statements that few can remember while others don't have one at all. The mission statement does not describe what the fire department does; most people already have an idea of that. The mission statement better describes how we do what we do. Most people can guess that the mission of the fire academy is to train firefighters. What makes the academy special is the way training is done. For example, look at the difference in the two mission statements that follow:

> *The mission of the Hanover Fire Academy is to deliver high quality, skills-based fire and EMS training to all members of the Hanover County Fire Department; to include recruits, firefighters, company officers, chief officers and instructors, in an effort to enhance volunteer recruitment and retention efforts.*

This mission statement does a good job of identifying what the mission is and who is trained, but only vaguely alludes to how it is done. The goal is to tell recruits and the public something they don't already know. A more accurate mission statement might be:

> *The mission of the Hanover Fire Academy is to provide all firefighters the skills they need to survive and protect others while demonstrating the principles and values that represent the proud history of the fire service.*

Though this mission statement is broader, it better describes how the training is delivered. It immediately instills a sense of pride and honor among those who read it. It is also short enough to be memorized and recited as the battle cry of the instructor staff. Each instructor must know the mission and be able to recite it on demand. This sets an example of discipline for the students and keeps the instructors focused on the mission, rather than other distractions.

When volunteer recruits join the fire service, their only impression of the fire service is what they've seen on television or through friends and relatives. They have little way of knowing the rich sense of tradition and honor that is the backbone of the service. Incorporating an honor statement is a way to tell the recruits what honor means to the fire service and, more specifically, to your department.

Honor statements, as mentioned in chapter 1, should be short statements that are memorized by both instructors and students. They are shouted on demand or at the completion of a particular task. For example, when running SCBA donning drills, the students are instructed to yell their honor statement as a way to signify that they have completed the task. The clock doesn't stop until the honor statement is heard. Repeating the honor statement over the course of their basic training instills the message in their minds for the rest of their volunteer career. When they see classmates years later, they may look at each other and shout the honor statement.

The honor statement must describe the absolute core of a department's values. Summing up those values into three to four words may take some time, serious thought, and effort to get right. Working with the instructors as a group may help to identify an appropriate statement. Find something that captures the feeling of the group, for example, the Hanover Fire Academy uses "Brother first, duty always!" as its honor statement. This statement describes the relationship among firefighters as a family, constantly looking out for one another, while remaining diligently focused on serving the public. Take some time to revisit the mission statement to see if it really captures how you do what you do. Consider creating an honor statement to bring the sense of honor and duty to all members of the department. Instructors must set the pace to honor our proud traditions and heritage.

Student Requirements and Expectations

Initially, volunteer recruits may be intimidated by the size and scope of the volunteer academy program, and it is critical that the student be made aware of the requirements and expectations of the program as early as possible. Once the rumors are clarified and the program is spelled out, most find it to be a welcome challenge. The intent of the various requirements and expectations is to ensure the students success rather than facilitate their failure, and this must be pointed out by the training officer at the beginning of the program. Strict rules and guidelines allow those who are committed to professionalism the opportunity to excel by weeding out the disruptive or distractive students. Most appreciate the quality training environment created by incorporating professional standards into the program.

The best way to motivate students to exceed expectations is to identify them in advance. The bar is set and each student has a clear target for achievement. Student performance expectations lists are developed that specify what is expected of the student at all times. The lists describe the type of firefighter that has developed over a long history of excellence and tradition in the fire service (Fig. 4–2). Expectations in 10 different categories are detailed to develop a description of what a firefighter should be. Each expectation is reviewed and explained to every volunteer. This gives them a chance to ask questions and ensures that they have a full understanding of what is expected.

Student Performance Expectation Categories

Judgement
Communications
Teamwork
Problem Solving
Initiative and Motivation
Work Ethic
Adaptability and Stress Management
Integrity
Community Awareness
Interpersonal Skills

Fig. 4–2 Student Performance Expectations (Courtesy Aurora, CO Fire Academy)

Judgment

Firefighters must develop sound judgment abilities early on in their years of service. New volunteers must come to realize that emergency services decisions can mean the difference between life and death. These decisions must be approached with safety and practicality in mind. Volunteer recruits are taught that decision-making is a privilege and honor that should be cherished and respected. This is done by initially limiting the number of decisions they are allowed to make. During the initial phases of training, every move is carefully supervised by the instructor. As the class progresses, the student is slowly allowed to participate in the decision-making process. The more the recruits demonstrate sound judgment, the more decisions they are allowed to make. This approach may be seen as militaristic by some, and that is correct. This process has proven to be highly effective in the armed services and fire service alike. Bringing it to the volunteer fire service must be done carefully. Pointing back to the mission and explaining to the department elders why this is necessary will help the training officer find success in this task.

Communication

Communications is listed as one of the highest dissatisfiers in the volunteer fire service. It is critical that the volunteer recruit learn to communicate effectively to lay a solid foundation for the future. New volunteer recruits are encouraged to carry themselves in a manner that is confident and professional at all times. Often, it is not what is said by the firefighter, but *how* it is said that causes problems, both internally and with the public. A zero tolerance policy for inappropriate behavior must be exercised by the instructor to ensure that such behavior is weeded out. Recruits must remember they represent the fire service 24 hours a day. In some cases, instructors have been known to correct communication troubles away from class while at the fire station. A hard approach to effective communication will pay off with a well-mannered professional firefighter in the end.

Teamwork

Seldom are there individual failures in the fire service. More often, such failures result from a failure on the part of a team. Unfortunately, in the fire service, such failures may result in serious injury or death. One of the key points in the basic program is that there are no individual failures. A student failing a written exam is a failure on the part of the entire company. The company should have provided that student with assistance to ensure that all members are prepared to be successful. The same approach applies to discipline. If one student is out of line, the whole company is punished, building a rapid sense of teamwork and account-ability for one's actions. Using these basic approaches to team building lays the groundwork for more advanced team drills, such as hose line advances and building searches. The only way the nozzle man can operate the nozzle to extinguish the fire is if the other team members get the nozzle and nozzle operator to the seat of the fire. Teamwork is vital to the success of any firefighting effort.

Problem solving

Historically, the fire service is the great problem solver. Finding creative ways to address problems is commonplace in the modern fire service, especially with the additions of technical rescue and hazardous materials. New volunteers are encouraged to address problems at their root, rather than the symptoms of the problem that lie on the periphery. Instructors often address initial complaints by talking the student into identifying root problems. For example, a company's inability to perform a practical skill may be symptomatic of an underlying problem. Until that root problem is identified and addressed, the symptomatic problems will continue to plague the group. Instructors must take care to lead the companies in careful problem-solving discussions to develop this necessary skill.

Initiative and motivation

When placed in a life or death situation, firefighters must be motivated to go the extra mile to get the job done. This same type of initiative is required on a daily basis to serve in a professional manner. Volunteers must be self-motivated to meet the various demands placed on them. Students who demonstrate initiative and self-motivation are rewarded, while those who require constant supervision pay the price of having the instructor watching them at all times.

Developing a motivated firefighter plants the seeds of future success for the volunteer department. Volunteer memberships that lack motivation and initiative often find it difficult to raise the funds necessary to survive as a modern volunteer department. Instructors must lead by example by demonstrating high levels of initiative and a desire to see a job done in a timely and professional manner.

Work ethic

Some say there is a diminishing work ethic plaguing the fire service. The next generation of firefighters somehow seems to lack the same motivation demonstrated by the previous generation. This may be true to some extent, but it is largely the fault of the generation making the complaints. New firefighters will mimic the same work ethic that is demonstrated as part of the culture of the department. Not the work ethic identified by the chief and the senior officers; rather, the work ethic demonstrated by veteran firefighters when no one else is around. Every firefighter is responsible for ensuring the next generation of firefighters knows how the job should be done, both on the fireground and in the station.

With staffing for both volunteer and career departments dwindling, departments must stretch the work force thinner than ever. Some volunteer firefighters seem to have the misconception that a working fire means they respond to the scene, use one tank of air, and then retire to the rehab area for the rest of the night for hot dogs and hamburgers. This overworks the few who are motivated to work and creates safety risks due to fatigue. In training, students are conditioned over time to work on a two-tank rotation in an effort to maximize the ability of the workforce. This may conflict with some opinions on emergency rehabilitation, but is a practical tactic on the modern fireground.

Adaptability and stress management

Emergency scenes can be some of the most stressful situations experienced in a human lifetime. It is imperative that volunteers learn to manage stress in all types of situations to ensure that they behave in a professional manner. Firefighters must learn to not judge citizens based on lifestyle, religion, or ethnic background. A classic tradition of the fire service is to provide quality service, regardless of the caller. Firefighters must be able to adapt to any situation and continue to perform in a manner consistent with this proud tradition.

Integrity

Integrity is the heading of a category that describes the key personal traits necessary to earn the title of firefighter. Our citizens are the true judges of what a firefighter should be, and they have developed their perceptions from our long-standing traditions and performance. This perception describes a firefighter who does what is ethical and moral; someone who is honest and dependable, highly competent, and motivated while demonstrating a heroic courage at all times. New volunteers must understand this expectation and realize that the fire service cannot afford to accept anything less.

The fire service possesses a public trust not found in other public service areas. Citizens invite firefighters into the most personal settings to care for their loved ones or solve their problems. This requires the highest level of trust and respect. Firefighters have an inherent responsibility to demonstrate integrity at all times. Who else is invited into someone's bedroom at three in the morning by a wife, still in her nightgown, to check on her husband who is feeling ill or who is injured? Explaining this special trust will help the new volunteer understand how sacred and special the fire service holds this honor.

Community awareness and interpersonal skills

Understanding the community is vital to ensuring the survival of the volunteer fire service. Volunteer leaders must constantly keep their fingers on the pulse of the citizenry to ensure that they understand the various beliefs and traditions found in the modern community. New volunteers must understand that they are not to judge those they serve for their beliefs or situations; they simply help, no matter what. This can be a difficult transition for some volunteer recruits who are used to speaking their minds. The things they say, even as a recruit can have a devastating impact on the volunteer department when communicated inappropriately. Negative communications can derail fund-raising campaigns and greatly damage the volunteer department's image and respect. Once the public loses trust in its fire department, it is extremely difficult to recover.

Effective interpersonal skills are the key to ensuring appropriate communication exchanges, with both the public and in-house. Volunteer recruits are expected to say, "yes sir" and "no ma'am" when addressing officers and members of the public. Their demeanor should be professional at all times. It is important that this be addressed, particularly given the changes in today's youth. For whatever reason, they may not have been taught appropriate manners, but this does not excuse them from the proud traditions of the fire service. The training officer must be prepared to provide whatever is lacking to ensure professionalism. Through repetitive conditioning, the recruit develops a professional manner that reflects positively on them and their department.

(Refer to the CD-ROM to view the Student Performance Expectations from the Aurora Fire Academy.)

The Code of Conduct

The code of conduct is vital to the volunteer fire academy process. Most volunteer training programs accept those who act in a manner that is less than professional and bend over backwards to see that they eventually get the training they need. The volunteer fire academy process uses a much different approach. Allowing students who lack the appropriate motivation and professionalism to continue in the program sends the message to the other students that there is no reward for excellence or integrity. If these negative factors continue to be a part of the system, they will soon spread to other students like a cancer, until the program either fails to be effective or collapses.

To prevent such events from occurring, there must be a means by which to remove volunteers from the program. The code of conduct spells out the deadly sins and the consequences that may result. Each student should read and sign a copy of the code that goes into their training file. This provides a valuable reference and documentation should a situation arise that requires disciplinary action. The signed document provides ironclad evidence that the volunteer was aware of the rules and the consequences for violating them, making it much easier to remove them from the program without damaging political clashes.

Each code of conduct must possess some basic categories to ensure that the training environment is professional and efficient.

Chain of command

The new volunteer recruit will likely have little or no knowledge of the chain of command unless they have prior emergency service or military experience. Each volunteer recruit must understand the management system used in the fire service and why it is important that it be practiced in all situations. The chain of command expectation should be based on the premise that all firefighters answer to one supervisor and that there is a process for doing business in the fire service, much as there is in their normal jobs. Practicing this concept in normal conditions will help prepare the volunteer for operating under emergency conditions when lives may be at stake. Using this discipline in the training environment will enable them to practice in both types of conditions to ensure that they will perform appropriately in the fire station. Instructors can use situations that arise during the class as teaching tools to demonstrate the appropriate use of the chain of command. Proper uses should be pointed out and applauded while inappropriate uses should be corrected immediately.

Effective use of the chain of command may also help to limit the amount of political turmoil that can be created by a disgruntled student. Those who may not be happy with learning good discipline, responsibility, and morals may attempt to use the political environment to shake up the system by spreading rumors through their volunteer companies or officers. By ensuring the student has a method of communicating ideas or complaints regarding the training process into the chain of command, this practice can be prevented or at least curtailed. Students who attempt to go around the system, attempting to exempt themselves from the high expectations of responsibility can easily be removed from the program because of violations to the code of conduct.

Respect

Respect is a two-way street. This category describes what is required from both the student and the instructor. Those who fail to show due respect are removed from the program. This zero tolerance approach helps to ensure an appropriate training environment that is professional and courteous.

Teamwork and effort

Physical performance is vital to safe fireground operations. This expectation is designed to weed out those who hoped to stay in the back of the pack, doing only what is absolutely necessary to get by. New volunteers must be prepared to give their best effort when on the training ground. Those who slack or fall behind are not tolerated. Recruits are also required to work as a team and must work in all positions during training. This policy helps to ensure that only those students who have demonstrated the ability required of a firefighter will graduate from the program.

Attendance

With every school that passes, like clockwork someone will ask, "How many classes can I miss?" Students must be reminded that firefighting is very dangerous and more than 100 firefighters die in the line of duty each year, the largest part of them volunteer. Each class in the training program covers a little piece of firefighting information that may save someone's life one day. One hundred percent attendance is mandatory!

Unfortunately, attending every class as scheduled is a tall order in the volunteer fire service. With increased responsibilities at home, at work, and at the firehouse, it is likely that each student will have some sort of scheduling conflict during the program. Training officers must consider this likelihood when developing the training program. Though 100% attendance is required, students should be permitted to miss a few classes and make up the work with an approved instructor. Creating a form to document the makeup work and placing it in the student's file will allow some headroom to absorb the unique challenges faced by volunteers. The number of classes the student can miss should be limited, and a deadline must be set for the makeup work to be completed.

Cheating

This is typically one of the shortest sections in the code and usually includes a simple sentence that states that cheating will not be tolerated and results in the immediate dismissal from the program.

Insubordination

Insubordination is the willful disobedience of an order given by a superior officer or supervisor. Such failure to follow orders on the fireground can result in injury or death, making it vital that new volunteers understand how to accept and carry out orders. Firefighters must be trained to follow orders without question, starting on the training ground and progressing to the fireground to ensure firefighter and public safety.

Appearance

Appearance is another area not typically addressed in the volunteer training process. The appearance category in the code of conduct should describe what the department requires regarding uniform and dress. Most volunteers train in street clothes unless their department has the resources to provide a uniform; this has historically worked well. The training officer must determine whether requiring uniforms for the students is a feasible option given the students' specific resources. Some academy programs are supported by their volunteer organizations to provide new members uniforms or to charge tuition to cover the cost. To comply with the demands of a volunteer system, training officers often allow street clothes during weeknight classes, as students sometimes come straight from work to class, but require uniforms for weekend classes. This works well to demonstrate the importance of a professional uniform appearance, while allowing exceptions to accommodate volunteer scheduling constraints.

There are some simple appearance stipulations that don't cost money or resources and help to project a more professional appearance. Though this type of clothing is popular among our volunteer firefighters, prohibiting clothing that projects a less than professional image of the fire service or may be perceived as offensive will help to bridge the gap between uniforms and street clothes. Allowing only approved fire department logos to be on the clothing, the training officer can do much to enhance the image of the department while avoiding a potentially offensive situation among the students or public.

Another difficult area to manage in regard to clothing is what can happen during live fire evolutions. The appearance of the sports bra has brought a completely new aspect to the training environment. It is understandable that volunteers will want to shed their wet shirts after exiting a hot training exercise and the volunteer training officer must be prepared to provide a means to do so appropriately. The appearance policy should describe appropriate clothing for rehab, and include when and where it is appropriate to change shirts for both men and women.

Substance abuse, discrimination, and harassment

These three categories address issues that can disrupt and even shut down a volunteer fire department. Each issue must be addressed by explaining exactly what is permissible and what is not and should be approved by legal counsel. Some volunteer companies have legal representation on retainer for such issues, while others do not. The volunteer training officer should work with the chief sponsor to ensure that the program is protected from any legal issues that may arise.

Performance expectations

This section adds weight to the performance expectations described earlier in this chapter. A simple statement saying that students must adhere to the fire academy performance expectations provides the training officer the ability to remove students from the program who demonstrate a substandard performance.

Accountability

This behavior reinforces the concepts described in both the student performance expectations and the code of conduct. Students are accountable for their own actions and their success in the training program is a direct reflection of their commitment to professionalism.

Discipline Policy and Prescription Agreement

Most volunteer training programs do not have a formal policy to address discipline. Most situations are handled at the instructor's discretion, or simply not handled at all. This makes it difficult for the instructor to enforce any conduct or performance requirements that may exist and gives the troublemaker student the upper hand when dealing with new or inexperienced instructors. The volunteer training officer must have a mechanism to exercise discipline when warranted for violation of the code of conduct. This is done based on the standard three-strikes rule commonly used in business. Establishing the guidelines for discipline provides both the instructor and the student clear direction for situations that are less than ideal.

Formal discipline and field discipline are the two main categories addressed in the company's discipline policy. Each plays a vital role in ensuring high standards of professionalism in the volunteer training program.

Field discipline

Field discipline is used to enforce minor corrections in behavior that may be required on the training ground. It is intended to serve as a motivator for professional behavior and is more symbolic than actually reprimanding. In most cases, once students realize that there will be consequences for inappropriate behavior, they cease to act in an unprofessional manner. Once the discipline is used a few times, it becomes a rare necessity for ensuring professionalism. Examples of behavior that require discipline include tardiness, lack of company accountability, being unprepared for practical stations, failing to provide one's best effort, etc.

Generally, the students vote as a group to decide what form of field discipline will be used for the class. Pushups are recommended, as they are easy to do and require little time to complete. Sit-ups or other calisthenics are other options for the students. As an alternative to physical activity, the students are offered the choice of writing a research paper, but this option is seldom selected.

Teambuilding is incorporated into the discipline process because one student should not be disciplined by himself or herself. When a student demonstrates behavior that warrants discipline, that student's entire company is assigned the discipline activity and they perform it as a group. This teaches the company that they are responsible for each other in all facets of the training. All successes and failures are a direct reflection of the team and they share both the ups and the downs together.

Formal discipline

Formal discipline addresses the more serious offenses that may result in a student's immediate removal from the program and outlines the steps that such discipline is carried out. For first offenses, the student is assigned field discipline. For second offenses, a prescription agreement is completed. Third offenses result in the student's removal from the program. It is vital that this be explained to each student, and each must sign a form stating that they understand the process.

Prescription agreement

The prescription agreement serves as a counseling form that outlines a process by which students can correct inappropriate behavior or performance. The form indicates exactly what violations have taken place or in what skill the student lacks proficiency. Since the instructor's goal is to help the student succeed, one section should indicate what actions the staff has taken and will take to offer the student every chance to excel. This statement will provide a summary of what has taken place and what the staff is prepared to do to help correct the problem.

The students are then required to provide an action plan that states how they plan to correct the problem. This plan should include a systematic description of how they plan to improve in the particular area. Benchmarks are established with dates or deadlines for the problem to be corrected. This procedure causes a future decision for dismissal to be based on the student's actions and places the ball entirely in their court. The form is signed by the student, the instructor involved, and the program coordinator and placed in the student's file.

Evaluation and Testing Policy

Performance evaluations are commonplace in the modern workforce, but seem to be a rare occurrence in the volunteer fire service. Without input from the instructors, volunteers have no way to know how they are doing in the training program. This omission can cause students who have the potential to be great firefighters to go unnoticed and those who are failing the program to fall by the wayside without special attention. Students should get regular feedback to ensure that they are on the right track and to set goals for the future.

Evaluations should take place as often as time and staffing permit. A minimum of bimonthly should be used for basic training programs. This schedule will provide the student feedback at least three times over the course of the training. A monthly schedule for evaluations is optimal to ensure the students stay on the right track.

The students are evaluated with a satisfactory/unsatisfactory rating on six categories. Each category outlines specifics as to what is acceptable performance and what is not. A key component of the evaluation is that students may not have any open prescription agreements to receive a satisfactory rating. The categories are academics, drill ground performance, performance expectations, physical fitness, equipment maintenance, and personal appearance. Each category receives an individual rating. Students with unsatisfactory ratings receive a prescription agreement that outlines how they plan to improve their performance using the next evaluation as a benchmark.

This method of evaluation provides the student a clear understanding of what is expected and how they can ensure that they perform to standards. The prescription agreement, with the code of conduct and performance expectation, enable the instructor some flexibility by stating that the student *may* be removed from the program. This is important when working with volunteers, as there will always be those special exceptions.

Testing

Taking written and practical tests are part of becoming a firefighter. Testing provides evaluation of both the students' performance and the effectiveness of the training program. Students often complain of the difficulty in taking the standard fire service test because of anxiety and stress associated with testing and poor test-taking skills. To better prepare students for the reality of such testing, volunteers are administered a written test with almost every class session and practical testing throughout the program.

Basic training curriculums often come with prewritten tests that can be administered at the end of each class session and software packages can be purchased to generate customized tests with the stroke of a key. Scores are maintained in a spreadsheet that determines individual, company, and class averages. Students must maintain a minimum 70% average to avoid being issued a prescription agreement.

Such testing facilitates increased reading of the text, which only improves the student's retention of the subject. This practice also conditions the student to better take tests in general, helping them to learn the nuances of test taking and techniques they can use to earn higher scores through practice and knowledge of test-writing skills. The extra effort will serve them well when faced with certification tests that used to cause high levels of anxiety.

Most firefighters must successfully demonstrate a handful of skills to gain certification at the end of the training program. Realistically, many skills are important to fighting fire safely. To ensure that no one slips through the cracks, skills testing is done periodically throughout the program. This provides the student practice in taking such skills tests and enables the instructor an opportunity to evaluate the student's abilities with much more scrutiny.

Company Assignments

In most cases, volunteers will come to class the first night seated by department, company, or local clique. These separations only fuel the political distractions that plague many volunteer departments. To promote teamwork and diversity, students should be assigned to training companies. Care should be taken to ensure that the class is mixed so that no one is with someone from their normal group. This is sometimes met with light resistance, but the gripes usually subside as people realize that they are there for the same reasons.

Companies are limited to groups of three to six people and are given alpha designations rather than the traditional numbers used in the fire service. This helps to separate the class from the local politics and allows the student to focus on training. Standard company designations are Alpha, Bravo, Charlie, Delta, Echo, and Foxtrot companies. Each company receives an accountability icon that they are responsible for at all times while training. The company icon is presented to the instructor with each skill station and an accountability report is given. Instructors will try to capture the company icon, which will warrant field discipline. This provides for some friendly competition between the instructor and students while fostering the importance of company accountability.

House Duties

House duties are a vital part of fire station life. With the diversity found in the modern volunteer applicant pool, it has become more important to include basic house duty training as part of the firefighter training process. Students must be taught how to mop floors and clean restrooms by the training staff rather than their parents. To facilitate this need, students rotate on house duties to keep the training facility clean. They are assigned by company on a rotating basis to clean the facility after class. This provides them some basic skills in cleaning and fosters a professional approach to taking pride in the cleanliness of their work areas.

Instructors must be sure to provide cleaning supplies and instruction as to what is expected when cleaning the facility. The funding for such supplies should be included in the training budget and the cleaning schedule should be posted in the classroom if appropriate. This simple addition to the training program will have a great impact on the volunteer department as it promotes pride in both the building and equipment.

Chapter Review Questions

1. Contrast the benefits of using a single academy coordinator vs. an academy board of directors.

2. The mission statement is designed to tell what two things?

3. What student document lists 10 different areas of performance in which students must excel?

4. Describe the two forms of discipline used in the fire academy process.

5. What document is used when students fail to meet expectations?

6. Why are company assignments made using alpha designations?

Answers

1. When using a single academy instructor, decisions can be made quickly, but the entire burden must be shouldered by one person. When using the board of directors, the burden is shared among the group, but rulings and direction comes slower, as the group must be assembled to make a decision.

2. What the department does and how it is done.

3. The performance expectations.

4. Field discipline is used to make minor corrections in behavior that may be required on the fireground and usually consists of pushups or other exercise. Formal discipline addresses more serious offenses that may result in the student's immediate removal from the program. The process is outlined step-by-step, including a first and second warning, plus final dismissal from the academy.

5. The prescription agreement.

6. To avoid any political associations with company affiliations from outside the program, as most volunteer fire companies carry a numeric designation. This helps break down the walls among students from different departments.

5

The Academy, Part 2:
The Key Components

Once the core components are in place for the basic training program, the volunteer training officer can choose which of the key components of the academy program best suit their department. Each brings a list of benefits that will enhance the volunteer training officer's efforts to recruit, train, and retain firefighters. This chapter provides an overview of each element. The training officer has the option to pick and choose among the options and use what best fits their department.

Drill and Ceremony Class

The drill and ceremony portion of the program is designed to be an eight-hour session that provides volunteers the basics in simple formations, honoring the American flag, honor guard, and funeral details. The need for such training stemmed from observing several volunteer departments on funeral detail. It was noticed that many of the volunteers present were unable to carry out the details in a way that honored both the fallen brother and the fire service. With some basic education and practice, such embarrassments can easily be avoided.

A key to implementing such a class is to ensure that the focus remains on the core mission elements. Frivolous marching and standing in formation does not help the department meet its goals and will cause unneeded political pressure. Be sure to constantly remind the class that the reason for such training is to be able to pay tribute to a fallen brother or sister in a manner that honors their incredible sacrifice. These skills are also used to honor the American flag. This tradition has taken on new significance since the tragic events of September 2001 that took the lives of more firefighters than any other single event in modern history. Those firefighters died serving their country and that sacrifice and commitment must be remembered and honored at every chance.

The class also does much to promote teamwork and discipline because it is based on the same training used in the modern military. Students first start to learn how to function as a company doing simple tasks, such as standing in formation and simple marching. This simple team building will culminate in complex live fire evolutions at the end of the program (Fig. 5–1).

Fig. 5–1 Students Learn to Stand in Basic Formation During Drill and Ceremony Training (Courtesy of Hanover Country Fire Training Academy)

Finding the instructors

It is difficult for someone who has no prior military experience to implement such a course. The subject matter requires an intimate knowledge of procedure and protocol common to veterans of the military service, making them first choice for establishing an instructor base. For volunteer fire service applications, examine your instructor pool for members who have such experience. If no such person exists inside the department, go to local law enforcement for help. Pervious military experience is very common in the criminal justice field, and drill and ceremony is often practiced as part of the daily routine. If such resources aren't available, contact local civic organizations such as the Veterans of Foreign Wars (VFW) for help.

Work with the available resources to create a written procedure for dealing with the basic ceremonies of raising the American flag, called posting colors, and providing honor guards and funeral details. Once the program is established, the current instructor pool must be trained first. Call an instructor training session on a weekend to ensure that all instructors have time to master the skills. This is especially important for departments that must rely on outside help to develop the program and their experts in the field will not be available at class time to provide leadership. Once the training officer is confident in the instructor pool's ability to deliver the program, the session can be added to the course schedule.

Daily use of the skills

The drill and ceremony program should be scheduled as close to the front of the basic training as possible. This plan will get the students outside and working as a team quickly. The program will help break the ice among the students and open a dialogue of teamwork. Once the initial training is delivered, students should practice their skills with every weekend class. Instead of sitting in the class waiting for roll call on Saturday morning, they are in formation awaiting the instructor to call them to attention. Each company leader must account for each of their assigned students. Companies rotate through the honor of posting colors, and company leaders each get a turn at being in charge of the entire class during the ceremony. This simple practice helps to foster discipline and honor in the volunteer firefighter.

Other components of the drill and ceremony process include paying honor to high-ranking members of the department or visitors who may be present at the training location. The training officer should determine at what rank the students should stop what they are doing and announce the presence of a senior officer. Often this includes officers above the rank of captain and typically the word "chief" is somewhere in their title. When such an officer enters the room, the first company leader to spot them announces, "Division Chief Smith in quarters!" This prompts the class to snap to attention until the senior officer entering the room offers direction. The entering officer will often order the class as they were, as they may be just paying the class a visit or passing through. In some cases, the chief officer may leave the class at attention if they have some business with the class that requires some form of reprimand. In either case, this practice goes far to impress the senior officers of the department and leaves an impression of discipline and professionalism.

Line of Duty Death Project

Every year an average of 100 firefighters die in the line of duty, the largest percentage of them being volunteer firefighters. It is the training officer's duty to ensure the hazards and threats associated with the fire service are made extremely clear to each volunteer firefighter as they move into the training process. The goal of the line of duty death project is to help the volunteers realize that the heroes who die on the job are just like them. The differences between life and death are often direct results of training, experience, and to some degree, luck.

In preparing to make the assignment, the training officer must do their homework to find cases that are appropriate for their particular department. Look for cases that happened in similar jurisdictions and resulted from traumatic circumstances, such as flashover, building collapse, or being caught and trapped. To find cases to choose from, go to the National Fire Academy web site, currently at:

URL: http://www.usfa.fema.gov/index.cfm
Keywords: Fallen Firefighter Database

This web site lists each line of duty death that occurs in the United States and offers a brief description of the circumstances. The tragedies are listed by year, allowing the training officer to choose current, almost up to the minute cases. Choose a case for each company, noting the name of the fallen hero, the name of the department, and date of the accident. This will serve as the baseline information for the class project.

Making the assignment

When making the assignment, care must be taken to reinforce one simple ground rule—at no time may anyone contact the family or immediate members of the department! There can be no doubt that this is forbidden and the consequences for violating this order must be severe. Since the cases used are very current, contacting the people who were close to the event can reopen wounds that they are working hard to heal and could easily be misunderstood or found to be offensive. Such contact with those left behind is absolutely forbidden.

What the students can do is contact the local media for newspaper accounts or a senior department representative to gather information about the event. The students are given a sort of script to use to contact these individuals that includes where they are from, and for what purpose the information will be used. Each company is tasked with preparing an eight page written report that provides a basic description of what happened and some general background information about the firefighter involved. A company presentation is assigned that must be a minimum of seven minutes and requires some form of participation from each company member. Students can use any manner of audio-visual support they feel necessary to help them relay the information to the class.

This assignment is made during the first few weeks of the training program and is not due until near the end of the six-month program. This provides the volunteer ample time to do the necessary research and juggle the study and practice requirements with their personal life. Care must be taken to periodically remind the class of the assignment to ensure that they pace their work over the six-month period.

Conducting the reporting session

The reporting session can be an emotionally charged meeting and fostering the appropriate attitude is vital to ensure that the students understand the intended meaning of the exercise. Using experienced instructors to lead the session will make a big difference in the outcome of the class session. Their experience will help students understand the aftermath of a line of duty death and the impact that goes beyond the obvious results.

Students are encouraged to think of their wives or husbands, their mothers and fathers, and the children they will miss seeing grow into adulthood. Most volunteers will think of these factors from their perspective, analyzing how the event will affect them personally. The point to be made is that their perspective does not matter in this situation, as they will be dead. They should instead be thinking of those left behind. Instead of thinking about how they will feel about missing their loved ones, they should think of how their loved ones will feel about missing them. Turn the tables in a Christmas Carol fashion, having the students imagine their family and friends attempting to recover from their death. Have them think of their grieving spouses, children, and parents. It is human nature for the inner dialogue to be selfish in such situations, thinking of only the personal perspective. Focusing the students' thoughts on others' perspective will cause a greater impact from the session.

After walking the class through the life-without-you scenario, the instructor can share any stories about their own personal loss in the fire service, as the intent of the project is to show the new volunteer that it can happen to them. Once the tone is set, the companies one by one make their presentations. Many students become teary-eyed as they hear the heartbreaking stories of losses that are real-life. The instructor should summarize each presentation to emphasize the key contributors to the deaths and the preventative measures that can be implemented to prevent such tragedies in the future.

To summarize the session, the lead instructor should point out the common factors that led to these tragedies. The instructor should emphasize that each fallen firefighter was just like the people in the room and that it can happen to anyone. Also, the reason for the many proud traditions of the fire service stem from honoring the huge sacrifices made by our brothers and sisters on any given day. The instructor can close the session by encouraging the class to stop for a moment and think of what is at stake when they enter a burning building. Each time they go in harm's way, they risk all that they hold dear in their family and friends.

Obstacle Course

Most volunteer training programs use only the minimum training requirements to provide SCBA training. True, this meets the general requirements for certification, but when it comes to such a key component in firefighter survival, the minimum requirements aren't good enough. Enhancing SCBA training and frequency of use will help provide the student with the confidence and skills they need to survive in life-threatening situations.

Identifying the need for the course

Aside from the firefighter survival benefits gained from the obstacle course, several other valuable benefits must be recognized. However, each volunteer training officer should evaluate the needs of their individual department and personnel to determine whether the obstacle course is feasible.

The obstacle course has several key benefits for the volunteer firefighter. First, it helps to build SCBA confidence through repetitive use. The best way to become familiar with the SCBA is to simply wear it. Running the course gets the student in the SCBA on a more regular basis, making them more confident and comfortable. Second, it enables practice in basic firefighter skills. By including common basic tasks routinely carried out on the fireground, the students quickly master the tricks of the trade for opening a hydrant, raising a ladder, or whatever skill is included in the course.

Secondly, with various complaints about the diminishing work ethic of the younger generation, the obstacle course provides a mechanism to develop stamina and determination that enhances the volunteers' desire to work. Some volunteer departments are blessed with an amazing support system that can put refreshments such as drinks, hot dogs, and hamburgers on the emergency scene almost as fast as firefighters can deliver water. Such services work well to rehab the troops on breaks and help improve moral. Unfortunately, this service also causes some volunteers to take more frequent trips to the rehab area and to spend more time there with each trip. By using the obstacle course to develop the volunteer's abilities, firefighters can improve to a two-tanks-take-a-break mentality as opposed to the one-tank-and-I'm-done approach that seems to grip some personnel. This two-bottle approach may not have the approval of specialists in physical fitness and rehabilitation, but is a real-life necessity on the fireground with ever growing personnel needs.

Creating a course

Some of the best obstacle courses are created through trial and error. The course should not be so demanding that it is unrealistic, but neither should it be so easy that just anyone can complete it. It should be designed to recreate physical exertion on the fireground in a manner that is as realistic as possible. Make a list of common fireground tasks that can be recreated in a safe manner, given the availability of resources in the training environment. No matter whether you have a regional training facility or a single firehouse, there are skills that can be used to create some type of course. With a little ingenuity, instructors can create courses that are challenging, and at the same time, cost effective.

Create a list of possible skill stations to develop the appropriate sequence of events that provides exercise for most of the main muscle groups. Figure 5–2 provides a sample of skill stations that make up a standard obstacle course. A pass/fail time is set at seven minutes, but students are not disciplined for failing to meet the minimum time requirement. Instead, they are encouraged to practice until they can be successful. The average time to complete the sample course is about four and a half minutes with some of the faster times being below the three-minute mark (Fig. 5–3).

Sample Obstacle Course

Stair climb with high-rise pack
Hose hoist
Descend stairs with high-rise pack
Open hydrant
Advance 200' of 2$^1/_2$" attack line
Walk/jog 150'
Crawl under obstacle
Drag 180-lb. mannequin 50'

Fig. 5–2 Sample Obstacle Course

Fig. 5–3 Student Participates in the Hose Hoist on the Obstacle Course

When determining exactly what stations to include in the course, have different instructors run the course to seek their input. Adding and removing skills to adjust the required effort will get the course to where it is challenging to the students, but not inappropriately difficult. Once a course is created that works for the department, it can be named to give it that sense of drama and suspense. Names like the Grinder or the Matrix are good to give the students something to brag about to their friends and coworkers.

The course is run at any available chance to get the students as much time on the course as possible. Since most practical evolutions for basic training under the academy format are run in skill-station format, it is easy to find time to run the course. For classes consisting of six companies, the course is scheduled as a double-skill rotation. A double-skill rotation consists of four firefighter skill rotations with one company assigned to each. The two remaining companies report for the obstacle course at the same time, one runs the course, and the other resets the course. Each company would rotate through the stations and the obstacle course throughout the day. The specifics of managing practical operations are discussed later in this chapter.

Fostering competition

Many students have found that the obstacle course portion of the academy program provides an unexpected sense of teamwork and competition. What many of them feared would complicate the challenges of volunteer training, rather created a new, more enhanced sense of motivation, teamwork, and cooperation. The instructor is the key in setting the tone for fostering such competition. Instructors must approach the course as a key training element that is serious, yet fun and exciting. They should be quick to point out the record times to beat and encourage the students to go for it.

The key to mastering the obstacle course is technique rather than brut force. Explaining this to the class is an important part of helping students gain passing times on the course and helping them gain confidence. During the initial runs on the course, the instructor should walk with the student, offering guidance and tips to help them work smarter, rather than harder. After becoming familiar with the techniques, students usually quickly cut their times and the race is on. Firefighters seem to be naturally competitive and the discussion among the class quickly turns to comparisons of times and challenges to do better.

The most unexpected benefit from the course is seeing students on their feet cheering one another. This reaction comes unsolicited by the instructors and is completely initiated by the students. Students having trouble completing the course have found their peers walking beside them, offering encouragement to move on.

This quickly helps to build the bonds of family and brotherhood. The relationships developed during the obstacle course help the students perform as a team later in the program as they move into the more advanced practical sessions.

Managing air-use data

Though the obstacle course is not based on any real scientific structure or data, it can provide an interesting look at how a student uses air to perform work and is an excellent teaching tool for demonstrating how to manage air use on the fireground (Fig. 5–4). This is done through logging each run of the course, capturing total air used in psi, and maintaining a record that plots each student's air use over the duration of the program. Provided in the included cd-rom, a sample air-use data form provides a place to log the students' times and air use on the course.

Fig. 5–4 Even When Facilities are Limited, Training Officers can Find Skills (here a volunteer crawls under an obstacle)

By tracking each use, students can see how they not only reduced their times in completing the course, but they also reduce the amount of air used to do the work. Tracking teaches them to manage their breathing to extend the work generated from one tank of air, a lesson difficult to learn early in the training process. This procedure can also identify students having difficulty using the SCBA by their inability to improve their statistics. This allows the instructor to work with them more closely to ensure that they are mastering this vital skill.

The obstacle course provides both the instructor and student a means of assessing the use of the SCBA, while fostering an invaluable bond between students as they embark on their careers in the volunteer fire service. The benefits of the course far outweigh the minor effort to establish, set up, and maintain this part of the program.

Family Day

With the average service life of a volunteer firefighter averaging five years, volunteer training officers spend huge amounts of time and resources training new volunteers, only for them to leave the service a few short years later. There are a wide variety of reasons firefighters stop volunteering, but a major reason is pressure from the family to be at home. This has prompted many volunteer fire departments to work to improve their methods of recruiting, training, and retaining firefighters. Finding a process that works well seems to be the holy grail of the volunteer fire service, with most struggling to keep their numbers at acceptable levels.

The volunteer training officer can have a great impact in retaining volunteers by incorporating the family into the training process. With the demanding schedules associated with firefighter training, most of the pressure on the student will come from the family, who does not understand why the volunteer is spending so much time away from home. Families often do not understand the honor and duty associated with fire service because what their loved one is doing is only an abstract notion to them. By incorporating the family into the training process, the training officer can help the student explain their endeavor in a way that applauds their commitment to duty and their community.

The training officer can incorporate the family into the training process by getting them out to see their hero in action on a periodic basis. This can be done by planning formal events for family participation or simply asking families to provide logistical support on occasion. Anything to get them out to see their loved one in action will help them to understand the gravity and importance of the commitment. It is important that these Family Days occur both near the beginning of the program and during the end. These sessions can be as involved or as simple as the training officers deem necessary for the department and class size (Fig. 5–5).

Fig. 5–5 The Family Should be a Major Component of the Volunteer Retention Effort

Scheduling Family Day

Two key days are vital to helping the student shoulder the burdens and pressures of home. Sometime during the fire hose section of the training in the first quarter of the program is an excellent day to get the families out to see the action. This gives the volunteer some relief early; about the time the serious complaining at home usually begins. Once the family members see that what the student is doing is meaningful and serious, they may be slightly more understanding regarding the family member's absence.

Secondly, toward the end of the program a full-blown family extravaganza should be planned to celebrate the student's achievement. This is a live fire exercise where students can demonstrate their abilities to ride the apparatus, stretch hose lines, attack fires, and make rescues. This session is planned for a time when students are finally in top shape and are near the end of the training program. The event is planned when families can get together to eat hamburgers and hot dogs and watch their loved ones save lives. This is a chance for the department to meet the families and establish relationships vital in retaining the volunteer over the long haul. Any time the training officer can get the families involved in the training process, it well help to ease the stress on the volunteer from the home front.

Providing the logistics

Depending on the scale of the particular Family Day planned, there are logistical considerations that are key to ensuring a successful event. The day must be carefully organized and planned to ensure that the families leave with a favorable impression of the fire department and the training operation. Failure to recognize the logistical needs of Family Day can cause it to backfire and create even more stress on the volunteer firefighter.

Obtaining a set of bleachers will help to ensure that everyone will have a place to sit and be comfortable. If the department lacks the funding to purchase bleachers, check with local little leagues or school sports organizations to see if old bleachers may be available. For informal training sites, such as academies that move from fire station to fire station, be sure to let the families know they will need to bring chairs to the training events. If funding allows, having tents on hand is beneficial, should inclement weather dampen the event. Also, ensure adequate parking is available and have public access to restrooms.

A major component of the Family Day event, as with most fire service functions, is food. It is important to provide time for the volunteers to mingle and meet each other's families and friends. This also provides time for the families to meet the instructor staff and get to know the fire service personally. Once the families have the faces to go with the names, they may be somewhat more understanding of the demanding training schedule.

The eating portion of the program can be as complex or as simple as available resources allow. For departments with ample facilities and funding, steak or spaghetti dinners can be served, while others may choose to serve hamburgers and hot dogs. The key to planning such events is to plan well in advance. Though students may give exact numbers as to how many of their family and friends plan to attend the event, plan for at least four family members per student. This will allow for the surprise family members who could make it at the last second. Once again, having a plan for rain will be beneficial, and dining indoors is always preferable to eating outside (Fig. 5–6).

Fig. 5–6 Providing Food and Shelter is a Major Component of Family Day

Documenting the event

A critical component of the Family Day process is to document the event through pictures and video. Be sure to get shots of the families sharing a meal together and plenty of action shots of each volunteer. Make sure to tell the students to get a picture of them in turnout gear posing with their families. The instructor should have a checklist with the names of each student to ensure no one is missed. These pictures will be a critical part of a multimedia presentation that is discussed in the graduation portion of this chapter.

Putting on the show

The highlight of what the families come to see is their hero in action, and the training officer must ensure that they get what they came for. Unlike normal training drills that are set up to create realistic fire conditions on the inside of the training structure, Family Day drills are designed to create fire conditions on the outside of the structure, providing an exciting show for the families. Setting the fires so that they are directly beside the windows will help to create a nice Hollywood effect to impress the families. Arranging for an announcer will add a sports event feel to the drill. The training officer should access the instructor pool for a staff member with the personality appropriate for such a duty. A public address system further adds excitement to the event and can include portable sound systems or the public address system on the fire apparatus. The announcer can call the drills, announcing which volunteer is doing what skill and provide background information to support the event (Fig. 5–7).

Fig. 5–7 Families Gather, Preparing to Watch Their Heroes in Action (an announcer calls the action)

Incorporating the family in the training process is relatively simple when compared to the benefit their involvement brings to the volunteer training equation. By targeting both the volunteer and their family for recruitment and retention, the training officer helps to expand the narrowing service life of the modern volunteer. Allowing the families to meet the instructor staff and see their loved one in action, gives the family members a much better understanding of the commitment and dedication necessary to be a volunteer firefighter.

The Formal Graduation

For years, volunteers have received their firefighter certifications quietly in the mail with little or no fanfare. This is a sad state of events, given they have just completed one of the greatest training challenges of their lives. This lack of fanfare does little to make the volunteer feel good about their achievement or encourage others to take on the task. A formal graduation ceremony offers all of the things lacking in the volunteer training program and, like Family Day, the benefits far outweigh the costs. Not only does it honor those who achieve the status of firefighter, it shows those yet to get the training what lies ahead and allows them to dream of walking across the stage to get their certificate.

Advanced planning considerations

Graduation ceremonies are formal events that require a great deal of planning. The most difficult ceremony to plan will be the first one, as the components must be created for the first time. Once a generic program is established, the training officer can use an academy timeline to set benchmarks on when to make the appropriate phone calls to arrange for the graduation ceremony. As the training officer develops the graduation ceremony, they establish contacts that can be used for future classes and most vendors will require only a simple phone call to place orders.

The first step in the planning process is to establish a date and retain a venue for the ceremony. This planning phase should be completed shortly after the training program starts to ensure plenty of time in case finding a location becomes difficult. Those who wait until the last minute risk having no venue at all. The date is scheduled shortly after the completion of the program, but allow time for any retesting that may be necessary. The size of the venue is based on the size of the class to determine how many people may attend the event. For classes of 15 or more, a high school auditorium may be required, as the facility must accommodate both the students' families and department members and officials. For smaller classes, a local fire station meeting room may be sufficient. It is important to verify the use of the facility in writing to ensure there will be no surprises on graduation night. Schools will usually require a signed agreement. For small venues, create your own agreement to provide documentation that the room has been reserved.

Once the date and venue are established, invitations must be developed. Invitations can be purchased from a stationary shop or created in-house on a computer. If purchasing the invitations, the local stationary vendor can advise you on what type of invitation is appropriate for the event and how to word the invitation itself. This provides a professional invitation that creates a first impression of the volunteer training program, which is extremely important. If creating an invitation in-house is more appropriate, make sure the invitation is at least as professional looking as a store-bought product. This can be achieved by purchasing an invitation package at an office supply store and using publication software to create the product.

Special guests and performers

For every graduation ceremony, there will be a host of different speakers and performers. The speakers are determined by the political environment and etiquette required for formal events. Making the contacts early will help ensure you get the first-pick speakers before their schedules fill up. Seek out speakers who bring a positive spin to the training experience that will excite the graduating class and the families alike. Don't hesitate to go outside the department or fire service to find a dynamic speaker. Local colleges and civic organizations may offer sources if no local fire service speakers can fill the bill. Other speakers who will be required will likely be the chief sponsor and other department dignitaries.

A nice touch to the graduation ceremony is the traditional bagpiper. This fire service tradition helps to set the tone for the ceremony as if to summon the spirits of those firefighters who have gone before. The local Emerald Society is an excellent place to start. If your locality has no such organization, search the local phone books, civic associations, and Internet to find a local piper in your area. Some will perform free while others will require payment. Usually whatever the cost, the bagpipers are well worth it to bring the appropriate feel to the ceremony (Fig. 5–8).

Fig. 5–8 Traditional Bagpipers Open Events in Honor of the Proud Traditions of the Fire Service (Courtesy of Hanover County Fire Training Academy)

Other performers may include local singers who perform the National Anthem or other appropriate songs. A key consideration when using vocalists is that you may have to supply audio support, depending on the size of the venue. Some departments purchase small sound systems to support such cases, while others may rent systems or avoid this component altogether. Make sure the vocalist will be valuable to the program, as those who do less than a good job will be the only thing everyone remembers about the ceremony.

Developing a program

The key to running a smooth graduation ceremony is to work out the details well in advance and practice as much as possible. Creating a formal program will help everyone involved know what is coming next and provides the audience a point of reference throughout the ceremony. Various publishing software is available to create a professional looking program. The program should outline the sequence of events, including the order of speakers, awards, and presentations. The program is also a good place to list academy sponsors and any other acknowledgements that are appropriate. Be sure that everyone who had a part in making the academy successful is recognized (Fig. 5–9).

Fig. 5–9 Formal Graduations Allow Volunteers to Celebrate their Achievements with Family and Friends (Courtesy of Hanover County Fire Training Academy)

Special presentations

Several different special presentations make the graduation ceremony a memorable experience for everyone. These presentations are simple and easy to implement with a little advanced planning. Some are in the form of awards and others are part of the ceremony itself. Each plays a valuable role in enhancing the students' feeling of achievement and accomplishment.

In the spirit of competition, the top graduating student should be given an award for their achievement. The winner of the award is determined by overall grade point average and skill ability on the fire training ground. Students are updated on who is in the lead periodically during the program to encourage students to strive to do better. The award can be a plaque documenting their achievement that is presented during the graduation ceremony. A plaque listing all of the previous top graduates can be placed at the training academy or fire station, giving future students something to set as a goal.

Another award unique to the academy program is the Honorable Duty Award. This award is given strictly to academy graduates who carry the academy values with them through their fire service career. It is given in response to some type of heroic or honorable action reported by a citizen or other member of the department. This award is cherished and only given when it is truly warranted. To determine who should receive the award, a committee may be appointed to review nominations for the award. The award may only be given every few years, depending on how much action a department sees.

Some departments assign a name to these awards honoring a previous or present member of the department; someone who demonstrated the values represented by the award throughout their career. Each department should evaluate whether they have a member in their history who deserves an award named after them. An example would be the John Doe Honorable Duty Award.

The class president elected at the beginning of the training program should be offered time to speak on behalf of the class. This time is used to reminisce about their experiences over the last six months and to poke some good fun at the instructors and various classmates. Some departments have traditions in which the class presents a plaque with a picture and names of the students and school number to be posted at the academy. These plaques are bought by the students and come unsolicited, making them even more special. At the Hanover (VA) Fire Academy, this tradition started out simple as the students presented gift certificates and hats to the instructors for their time and efforts. This practice gradually evolved into the traditional plaques that now line the halls of the fire academy.

Following the class president and presentation of the class plaque, a multimedia presentation can be shown, documenting the students' adventures over the last months. This is created by accumulating pictures over the course of the school and assembling them into a PowerPoint slide presentation. The training officer must ensure that they get pictures of the main training components throughout the program and save them for the graduation ceremony. Once the course is near its end, the training officer can sort the photos, looking for the best pictures to use at the graduation ceremony. Once selections are made, they can be set to music to create a unique snapshot of the highs and lows associated with the program.

The multimedia can be divided into two main sections: (1) capturing the bulk of the action with plenty of fire and sweat, and (2) capturing Family Day and the sacrifices made by both the volunteers and their families. The first action section is set to upbeat, aggressive music, while the family portion is set to a softer, gentler tune. The family portion should complement the photos from Family Day of the volunteers posing with their families in their turnout gear and mingling with the other families and staff.

To create a trouble-free multimedia presentation, a few tricks will help. First, determine what two songs will be used and get the exact length of each tune. Add the two times together and determine the total number of seconds. Divide the total number of seconds by the number of pictures in the presentation to determine how much time can be spent on each slide. This will help to create a polished presentation that coincides with the music.

There are two ways to insert the music into the program. First, the songs can be added as a custom animation, starting on a mouse click set in advance by using the *rehearse timings* feature in PowerPoint. Transfer the selected songs to one disc, but be sure to include one extra track. There is a glitch in PowerPoint that requires an extra track to be on the CD for it to play properly in this application. The second and easiest option is to play the CD in a regular CD player and start the music and PowerPoint presentation simultaneously. This gives the operator more control over sound quality and is easier to create.

Another added feature to the graduation ceremony that tops off the program is to provide refreshments after the event (Fig. 5–10). This is a simple addition featuring cake, ice cream, and punch. The volunteer ladies auxiliary or other support group is a good source of help for adding this feature to the program. Refreshments offer time for the graduates and their families to mingle one last time for photos and storytelling. This feature may be added through donations from the local community or may be paid out of the department budget. Approximately $200 should be allotted for classes of roughly 20 students.

Though the graduation ceremony requires detailed planning, the benefits derived from the event are well worth the effort. Providing rewards for both the volunteers and their families for their hard work will pay off in the future with increased recruitment and retention statistics.

Fig. 5–10 A Post-Graduation Reception Allows the Families to Mingle and Celebrate (Courtesy of Hanover County Fire Training Academy)

Remember the ideas presented are designed to be modular, allowing the training officer to implement each independently, or as a group. Some may choose to phase-in the options over time to reduce the perception change in the department. In any case, each option carries little, if any, associated costs and will do much to improve the quality of the training program. As the training officer transitions through the implementation phase of these options, they must also be prepared to reevaluate their approach to managing their practical operations, as they too will be affected.

Managing Practical Operations

One of the most challenging tasks facing the volunteer training officer is determining how to manage larger classes on the training ground. The goal is to keep everyone busy, making the most of the volunteers' training time. With classes of 20 or more, the training officer may have to take a different approach to keep all of the students working with minimal downtime.

The old days of one instructor with a class full of students are gone. The training officer must find additional instructors to make the practical evolutions effective. This will require extra work by the training officer to find or create enough help to run the training program. A good rule of thumb is to have one instructor per company in the class, plus a coordinator to oversee the training operation. For cases when the training officer is one instructor short, the coordinator may have to double as both coordinator and instructor.

In the planning phase, the training officer can determine exactly what skills need to be practiced and a logical sequence of events for the day. A written set of objectives can be created for each station outlining exactly which skills must be completed and what is an acceptable performance. The skill station sheet should also include a list of required equipment necessary to carry out the evolutions. Instructors should be assigned to a skill station and given time to familiarize themselves with the operation. When the companies report to the skill station, the instructor receives the accountability report and verifies to the company leader that all company members are accounted for. They may then proceed with the assigned skill station. Figure 5–11 provides a sample skills schedule that can be applied to almost any practical training session.

Rotation	Company Assignments					
Skill Station #1	Alpha	Foxtrot	Echo	Delta	Charlie	Bravo
Skill Station #2	Bravo	Alpha	Foxtrot	Echo	Delta	Charlie
Skill Station #3	Charlie	Bravo	Alpha	Foxtrot	Echo	Delta
Skill Station #4	Delta	Charlie	Bravo	Alpha	Foxtrot	Echo
Skill Station #5	Echo	Delta	Charlie	Bravo	Alpha	Foxtrot
Skill Station #6	Foxtrot	Echo	Delta	Charlie	Bravo	Alpha

Fig. 5–11 A Skill Matrix Can be Created for as Many Skill Stations as Necessary

The focus for the skill coordinator is to ensure that all safety requirements are followed and that things move smoothly during the rotations. In a best-case scenario, the coordinator can appoint a safety officer for the operation. With many departments struggling to keep instructors in front of the class, an instructor designated specifically to safety may be a long shot. The coordinator should also check with each instructor to ensure that they have everything they need and that the skill objectives are being met properly. Students unable to complete a skill are sent to the coordinator so that the incident can be documented and forwarded to the class coordinator.

It is vital that the skill coordinator ensure that the skill rotations are kept to the predetermined time allotments. Typically, skill stations are limited to 45–50 minutes with a 10-minute break in between. This helps keep the day on schedule and prevents training sessions from going beyond their scheduled time. The coordinator should give instructors a 5-minute warning, informing them to prepare to wrap up their operation and prepare the company for transfer. The training officer must remember that the volunteer has many other commitments besides the fire department, and care must be taken to respect those obligations. Finishing class on time will help to ease the stress on the student in their personal life.

Another important component of the coordinator's duties is to ensure that anyone from outside the department or class training with the group is in compliance with *NFPA 1403, Standard for Live Fire Training Evolutions.* In cases where the coordinator is familiar with all of the students involved, it is easier for the training officer to document this requirement. When working with unfamiliar students, the coordinator can have each student sign a written form stating their compliance with the NFPA standard. This form should be kept on file to protect both the instructor and the department.

Chapter Review Questions

1. In what two major ways is drill and ceremony used in the daily application of skills?

2. What is the golden rule of the line of duty death project that the training officer must ensure students never break?

3. What two key benefits come from using a SCBA obstacle course?

4. Name one of the key objectives for the lead instructor when managing practical operations.

Answers

1. The posting of colors and announcing the presence of chief officers.

2. Contacting the family of the fallen brother or sister.

3. The obstacle course enhances the student's proficiency with SCBA and builds the student's work ethic.

4. Ensure that skill station rotations stay on schedule.

6

In-Service Training Options

One of the greatest challenges facing today's volunteer training officer is finding ways to make in-service training meaningful and exciting for their volunteer firefighters. Such training should offer value to the department and help move the firefighters toward departmental goals, while offering topics that the volunteer firefighter finds interesting and challenging. This is a tall order, given the vast diversity of experience in the volunteer force. Incorporating differences in age, experience, and personal preference into one program creates a huge challenge to meet the expectations of everyone involved.

Politically, training officers are sometimes measured by their ability to provide training that appeals to the firefighter ranks, regardless of the other job duties they do well. Finding guidance in developing in-service training can be difficult. There is little information on the market providing ideas for such training, and communication between volunteer departments has been historically limited. As with most training processes, the training officer must develop a systems approach to in-service training that is somewhat self-sufficient and rejuvenating. Many training officers realize that in-service training is only a small part of the overall training system. Senior members of the department quickly forget about the initial training and other components that the training officer must provide all new volunteers and form their opinion of the training program solely from the in-service portion they personally experience. This makes in-service training a potential sore spot for the volunteer training officer.

The basics of in-service training were briefly discussed in chapter 4 in the section on building a training system. The current chapter will look in depth into how the training officer can develop a training program that is based on needs assessments and performance evaluations to provide training challenging to the volunteer.

Education–Drilling–Training

Many training officers view in-service training as a vast hunger they are unable to satisfy for their membership no matter how hard they try. The in-service issue can be broken into chunks that are more manageable by looking at the various phases of the traditional training process. In-service training can be divided into three parts that can help the training officer achieve a more manageable span of control over the various components of their training system. Each has its own target audience and method of instruction that must be recognized and incorporated into the training system.

Education refers to the initial learning process that serves as the foundation for which new information can be applied. In the regular job market, a high school diploma, by and large, is considered the baseline education that employers expect from their employees, even though today's standards are quickly being raised to a college level education. In the volunteer fire service, education refers to the basic firefighter training, such as Firefighters I and II or completing the academy process.

Many volunteer training officers find it difficult to get past the initial education process because of high turnover rates in the membership. Departments that focus on quality rather than quantity tend to see fewer turnovers, and, after a while, begin to get past the initial education phase. This requires a great deal of commitment from the department and training staff to ensure that high standards are maintained throughout all facets of the department, even amid occasional intense political pressure for mediocrity.

Drilling refers to the reinforcement process used to ensure that firefighters maintain current knowledge. Some drilling is done during the educational phase to ensure that students possess a minimum proficiency, but drilling can also be used by veteran members to maintain proficiency on rarely used skills. By establishing minimum performance standards, the training officer can use periodic drills to ensure that the department or company maintains that minimum proficiency. Go back to *NFPA 1410* to establish standards on the common fireground tasks. This provides firefighters a target to drill for and may even promote some friendly competition, a reliable motivator in the fire service.

The term *training* implies that new knowledge or skills will be added to a pre-existing knowledge base, building on previous experiences and abilities. For example, firefighters may receive training on new low-pressure nozzles that the department recently purchased. They already have an education on fire behavior, streams, and tactics; they are simply receiving training to supplement their prior knowledge.

Developing an in-service program that incorporates all three phases of the learning process will help the training officer develop a program that both meets the learner expectations for periodic training, and the department's goals for performance and quality. The program should include educational programs for new disciplines or technologies that impact the volunteer fire service, such as emergency medical services. Departments find themselves providing medical educations to their members to keep up with increased demands for medical service in the community. A drilling component should be in place to maintain proficiencies in the basics, while training is provided on new additions to old skills, such as thermal imaging. By developing a comprehensive in-service system, the training officer will have the ability to provide classes that are interesting to the volunteer and meaningful to the department.

In-Service Responsibilities

Understanding who is responsible for in-service programs will help place the workload where it belongs and bring efficiency to the program. Contrary to the beliefs of some in our volunteer system, the responsibility for in-service training does not lie solely with the volunteer training officer. This is a copout of responsibility on the part of the company officer to think that they are immune from all training responsibilities and all of the fault or glory goes to the training officer alone. The *learning handshake* identifies who bears the responsibility for the various levels of the in-service program (Fig. 6–1). The responsibility for each learning level shifts as it moves from the education to the training categories. Though everyone is ultimately responsible for their own abilities and education, it helps to understand where the priorities lie as to who should be primarily responsible for the production of in-service programs.

Learning Level	Responsibility
Education	Training staff—company officer
Drilling	Company officer—training staff
Training	Training officer—company officer

Fig. 6–1 In-Service Responsibilities

During the education level of learning, the responsibility lies primarily with the formal training staff or training officer to provide the baseline education. It is the company officer's responsibility to make education a priority within their company, and they may even assist in the delivery of the basic education program. As the focus shifts to the drilling level, the responsibility for in-service programs shifts to the company officer. The company officer knows the specific needs of both their crew and the response area. They are responsible to schedule such drills as part of the weekly routine and focus the program on areas that need practice. The responsibility for the training phase of learning is shared by both the training and company officers. The training officer works with the company officer to offer training that is responsive to the needs of the company, while incorporating the new information and skills necessary to meet departmental goals. Understanding the working relationship in advance will help all involved develop an in-service program that will be successful and interesting for the participants.

Once everyone is aware of their role in the in-service program, the first step is to refer to the needs assessment to identify any skill deficiencies that may exist in the current operational system. Some volunteer departments may have detailed operational guidelines for performance to measure their department's abilities, while others may need assistance in determining what is acceptable and a method to measure these abilities. Two NFPA standards offer guidance to meet this challenge. *NFPA 1410, Standard on Training for Initial Emergency Operations,* and *NFPA 1720, Standard for the Organization and Deployment of Fire Suppression Operations, Emergency Medical Operations, and Special Operations to the Public by Volunteer Departments* offer excellent guidance in evaluating volunteer department needs and help establish in-service training priorities based on actual company performance.

Using *NFPA 1410*

NFPA 1410 "specifies basic evolutions that can be adapted to local conditions and serves as a standard mechanism for the evaluation of minimum acceptable performance during training for initial fire suppression and rescue activities."[1] This standard offers a baseline assessment for the in-service training system. One of firefighters' greatest complaints is that training becomes boring or redundant. Using a baseline measurement to establish the in-service training priorities will help keep the training focused where it needs to be, on the essentials of firefighting.

High-risk/low-frequency training such as the *working fire* should serve as the cornerstone of the in-service program. *NFPA 1410* has established performance categories by which to measure departments. The categories listed are required performance for handline and master stream operation, sprinkler system support, and truck company operations. Each category specifically outlines the performance criteria and offers a sample evaluation method. Measuring these performance categories will help the training officer determine which areas of fireground operations should be given priority in relationship to structural firefighting.

The standard also includes 14 *ready-to-use* evolutions that work perfectly for the initial in-service drills. The standard provides diagrams and specific information to make it easy for the training officer to put these exercises into action. This guide provides the training officer up to a year's worth of in-service training options and provides an assessment of the department at the same time. Simply schedule a specific drill that is appropriate to the department for each drill until all areas of service provided by the department have been assessed.

Once the entire assessment has been completed, the training officer will have a much better idea of the abilities of the department and what training priorities are realistic and appropriate. After the initial assessment, the individual drills can be peppered into the training schedule to provide periodic evaluation and maintenance of skills.

Another unique benefit of using *NFPA 1410* is that it causes departments to look hard at how they do business on the fireground. The standard allows companies to work out the bugs in their tactics by having them apply their skills under the scrutiny of time. This will flush out the little things that would go unnoticed until the next big fire. Things as small as making sure that the couplings on the supply hose connect and disconnect smoothly will be identified as a potential problem that might delay a fire attack in a real situation. Such work enables companies to iron out the details that will ensure the attack is initiated smoothly when there is more on the line than the fastest time.

Using *NFPA 1720*

The requirements of *NFPA 1720* "address functions and outcomes of fire department emergency service delivery, response capabilities and resources."[2] It establishes a standard on fire suppression organization and operations that describes the responsibilities of the incident commander and emergency management system specifically for the volunteer setting. This information provides an excellent guide for the volunteer company drill, taking each part of the standard to develop a short two- to three-hour training session. This approach not only improves performance, but helps to ensure compliance with the professional standard as well.

Section 4.2 of the standard outlines the requirements for initial fire attack as a supplement to *NFPA 1410*. Volunteer training officers can provide drills to help the department with skill proficiency by meeting the standard as a goal. Section 4.3 describes requirements for mutual aid agreements, prompting multi-company training among neighboring departments. The standard also provides requirements for emergency medical services and special operations, such as hazardous materials response. By using these professional standards as a benchmark, the training officer can create training that is meaningful to the department by achieving specific training goals The training will also be interesting to the volunteer firefighter because it has a real-world effect on their performance.

Finding the Right Topics

Another key to in-service success is staying in touch with the interests of the company. These popular topics should be incorporated with the educational, drilling, and training components of the in-service system. Many training officers complain that they run out of ideas for in-service training, but when all of the system components are considered, it seems that these training officers just aren't looking in the right places. Keeping up with the popular trends of the day, new educational requirements, drill proficiencies, and training on advances in tools and equipment, there should be a variety of in-service topics from which to choose. Rather than looking for in-service programs to teach, the training officer should be challenged with how to fit all of the in-service opportunities into the schedule.

The various in-service options can be categorized to help make sense of the various opportunities. Sorting the ideas by single company- and multi-company-oriented ideas will help with logistics and planning to ensure that the program runs smoothly. Both are intermingled to keep the volunteers interested.

Setting Goals and Proficiency Standards

A key for success with in-service programs is to communicate the goals and expectations of the session upfront. This provides the volunteer a clear understanding of what is in store, making the less pleasant topics more tolerant. A simple statement of expectations, with a description of the acceptable performance will provide everyone a clear understanding of what is to be done. Figure 6–2 contains a sample single-company form that can be used to express expectations and document performance. It plainly states the objectives of the session and documents the skill proficiency for each student.

Hanover County Fire Training Academy
Firefighter In-Service Training

Skills Sheet

Topic: Ropes & Knots

Student Name: _____

Date: _____

Instructor Name: _____

Hours Training: _____

Pass	Fail	Skill #1
☐	☐	**Bowline Knot**
The firefighter shall demonstrate tying a bowline knot with a safety on a smoke ejector in preparation for hoisting. The firefighter shall verbally identify the working, standing, and running end of sections of the rope.

Pass	Fail	Skill #2
☐	☐	**Clove Hitch**
The firefighter shall demonstrate tying a clove hitch with a safety and half hitches on a pike pole, in preparation for hoisting. The firefighter shall verbally identify the working, standing, and running end of sections of the rope.

Pass	Fail	Skill #3
☐	☐	**Figure Eight on a Bight**
The firefighter shall demonstrate tying a figure eight on a bight in preparation for hoisting a ladder. The firefighter shall verbally identify the working, standing, and running end of sections of the rope.

Pass	Fail	Skill #4
☐	☐	**Handcuff Knot**
The firefighter shall demonstrate tying a handcuff knot in preparation for lifting a downed firefighter. The firefighter shall demonstrate proper knot placement on simulated victim and proper rope set-up for Rapid Intervention Team operations.

Comments: _____
_____
_____

Fig. 6–2 Sample In-Service Training Skills Form

Single-Company, In-Service Options

Single-company, in-service options play an important role in the training system because they are a delivery mechanism that can move small chunks of information throughout the entire department very quickly. For small volunteer departments, the single-company option is the primary in-service delivery method. In either case, the single-company option is the workhorse of the in-service program. Making single company drills as efficient as possible will help the training officer get the most for their training efforts.

The single-company option can be used for the delivery of topics that stand alone, or in pursuit of a larger goal. Stand-alone training consists of topics that serve as refresher courses or subjects that do not require a great deal of time to carry out. Figure 6–3 provides some sample single-company, in-service training topics. These topics are good in-service options because they enable the firefighter to refresh skills that they may not have used since their initial education. Each could be conducted on a drill night or on a weekend, depending on how in-depth the training officer wishes to go.

Single Company Training Topics

Blood-borne pathogens training updates
Automatic external defibrillator refresher training
Ropes and knots
Ladders
Ventilation
Confined space awareness
Salvage and overhaul
Hazardous materials refresher training
Forcible entry

Fig. 6–3 Varieties of In-Service Training Options are Available to Training Officers

One of the main benefits of single-company training is that the training officer can assign the training to a qualified instructor. Many company officers are qualified instructors and are a good choice for delivering the programs because they will be working with the crews on the fireground. With an in-service skill sheet (Fig. 6–2), the company officer can deliver the training, document the proficiencies, and return the completed forms to the training officer. This allows the volunteer training officer time to focus on the entire training system, rather than being mired down in the delivery itself.

Some training officers schedule small, single-company drills that culminate in a larger multi-company exercise. This unique process ties the various training subjects together and facilitates a nice mix of single- and multi-company training. Such events require advanced planning to ensure that all the pieces flow in the correct order and that the training is consistent with the end goal. For example, if the volunteer training officer knows that the company is planning a large multi-company evolution, such as burning an acquired structure in six months, he or she may incorporate the smaller training components that will be needed for the larger event into the in-service schedule ahead of time. With five months notice, refresher topics such as rural water supply, hose line advancement, forcible entry techniques, and residential search techniques can be added to the schedule. This enables all of the various preparatory topics to come together at the acquired structure. It allows the firefighter a chance to see how all of the various topics intertwine to play important roles on the fireground.

Whether for specific individual topics or in pursuit of a larger training goal, single-company, in-service training serves as the backbone of the department's ability to maintain proficiency on the fireground. The training officer must work closely with the company officer to ensure that such training is a priority and is meaningful to both the firefighter and the department.

Multi-Company Training

Multi-company training is a vital component of the in-service system because it gets volunteer companies together in a training environment. This is critical as few departments can handle a working fire without a little help from their friends. Multi-company training should place companies into situations in which they must work together that are consistent with the hazards in the local community. Such events should be scheduled periodically throughout the calendar year and in coordination with other single-company evolutions.

The volunteer training officer from the single-station department must develop relationships with the surrounding jurisdictions with which they have mutual aide agreements. Better yet, they should establish relationships with those who have no formal mutual aide agreement, but are in close enough proximity that they may work together during the fluke catastrophe. Simply getting companies together will help facilitate a better working relationship because they will already know the faces and maybe even the capabilities of the *other* guys.

Once the training officer has established contact and introduced everyone, the groups can start to discuss strategies and tactics and develop formal plans for working together. Establishing such a dialogue will facilitate a coordinated and efficient response to that once-in-a-lifetime situation that demands an appropriate response to save lives and property.

For the training officer in the larger, multi-company department, the challenge may be getting your own companies together on strategies and tactics. Most multi-company volunteer departments were at one time individual departments that were eventually consolidated into a county or protection district to handle increasing call loads and to save money. Many of the companies will tend to retain their individuality in spite of the department's efforts to make them seem as one. This sometimes strains the relationship among the companies and makes the training officer's job more challenging to say the least.

Once the training officer gets the department beyond the consolidation blues, they can focus on mastering multi-company tactics. Most fireground operations require multiple companies to carry out choreographed tasks with precision and accuracy. To facilitate such a performance, the training officer must focus their initial efforts on the basics to ensure that water gets to the fire. Depending on whether the department is rural, suburban, or urban, the training officer may start with water supply operations as a way to open a meaningful dialogue among the groups to determine the best method to deliver water, given the local environment. Moving discussion to practice will allow the various companies to work as a team and build valuable bonds critical to future response.

As the companies master the bread and butter operations, they move on to complex operations, such as hazardous materials response or technical rescue. Many localities have found success in providing technical rescue services by sharing the burden with their neighboring jurisdictions. No one locality could provide the funding, staffing, and training necessary to field a full technical rescue team, but by working together, they could share resources. This sharing of abilities and equipment allows the region to respond with the capabilities reminiscent of a Federal Emergency Management Agency (FEMA) task force.

The volunteer training officer can schedule multi-company drills periodically, in conjunction with the single-company training to create a program that is dynamic and exciting, yet in pursuit of company or department goals. With all of the various training options available, it seems more difficult to choose what to do first, rather than stumbling around looking for training topics.

Self-Study Options

As technology continues to speed up the pace of life, the training officer is provided more options to consider for in-service training. Computers offer a new option for firefighters to access training programs at their own pace, rather than gathering at the firehouse on the predetermined night for training. Advances in technology will not relieve the volunteer firefighter from attending hands-on training, but it may ease the time requirements for some lecture-based education and training. This may allow the volunteer firefighter to complete the education or training objectives on their own time, maybe even from home, depending on the level of technology available to the department. Such programs can even be used by relatively low-tech departments using a print-based program that affords the volunteer firefighter the same flexibility in scheduling.

Such programs typically require the firefighter to read the material and then prove comprehension by answering test questions to document their proficiency. In same cases, these types of programs can go beyond the boundaries of cognitive learning by incorporating self-study practical skills for simple training programs. This is possible by including very detailed skill sheets that can be completed and turned in to the training officer after completion.

The college systems have made good progress in incorporating such programs into their curriculum. Many have self-study courses that are either entirely print-based, or a combination of assigned reading and lecture. This is accomplished either by picking up class session videos or, in some cases, using the Internet to attend virtual class sessions. These tools allow the student to work according to their own schedule and submit evidence of their performance by a predetermined deadline. This unique method of education must be capitalized on by the volunteer training officer as a possible alternative method for meeting training requirements within the constraints of the volunteer schedule.

Resources for such educational formats include the local community college. Contact the nearest community college to find out what they are doing for distance education and what might be available to the local volunteer fire department as a resource. A common medium used by the community college system is software called *Blackboard* that allows students and instructors to communicate assignments and questions over the Internet. This system gives the student constant access to the instructor and training curriculum by enabling the student to view questions for discussion from both the instructor and other classmates. The program is expensive for most volunteer departments, but the training officer may be able to arrange for access to the medium by partnering with the college. For more on *Blackboard,* you should contact your local community college or go to www.blackboard.com for more information.

Using Outside Sources

Even the best fire service instructors can become boring after a while. To keep things interesting, the volunteer training officer should plan to incorporate fresh faces in the training calendar. This includes fire service professionals from other jurisdictions and experts from outside the fire service entirely.

There are several ways to approach getting outside instructors for in-service training. It may be interesting for the firefighters to hear how other departments address a common problem, either by a department similar to theirs, or by a department entirely different. Bringing a city firefighter out to the suburbs to discuss search and rescue may offer a unique perspective, not because the other department does it better, but because they simply may have come across situations that your department has not. Many times such training events are a learning experience for all parties involved, providing each a fresh perspective on an old topic.

In other cases, a neighboring department may have a guru in a particular topic that can be brought in at low or no cost. Most instructors tend to find a specialty or niche that they enjoy teaching and develop training programs that are far better than the average training curriculum. That instructor who enjoys teaching building construction may have constructed a miniature wood frame structure to scale that will amaze even the veteran volunteer firefighters, simply because of their passion and mastery of the topic.

Another way to incorporate outside resources into the in-service program is to step entirely out of the fire service box. It takes far more than water and a hose to make the average volunteer fire station run smoothly; it also takes things such as personnel, financial, and administrative management. Most jurisdictions have access to a variety of experts in all phases of fire station life who would be happy to come share their knowledge in their area of expertise. For example, a class in conflict resolution would be of great benefit to any fire department. There is likely a specialist who would provide training free of charge to help the volunteer fire department. The key is for the volunteer training officer to think *outside the box* to find new and interesting ways to provide in-service programs.

The volunteer training officer has a wide variety of in-service options readily available. Rather than complaining of a lack of training ideas, the training officer should find it difficult to get all of the options they wish to address into the training calendar. By using some advanced distance learning techniques, training officers can maximize the volunteer firefighter's ability to meet ever-growing training requirements while balancing the demands of their personal lives.

Chapter Review Questions

1. Define what education, drilling, and training mean when referring to in-service training programs.

2. Describe the elements of the learning handshake and outline who endures most of the responsibility for the various phases of learning.

3. What NFPA standard describes possible in-service options?

4. What is one of the main benefits to the training officer for using single-company, in-service training?

Answers

1. Education refers to the initial learning process that serves as the foundation for which new information can be applied. Drilling refers to the reinforcement process to ensure that firefighters maintain current knowledge. Training implies that new skills or knowledge will be added to a preexisting knowledge base.

2. The learning handshake refers to who bears the responsibility for education, drilling, and training. The training officer is primarily responsible for the education phase, while the company officer bears much of the responsibility for drilling. The training responsibility is shared between the training officer and company officer to ensure that firefighters receive the best training possible.

3. *NFPA 1410.*

4. The training officer can assign training to a qualified instructor.

Notes

[1] National Fire Protection Association, 2000, *NFPA 1410, Standard on Training for Initial Emergency Scene Operations* 1-1.2.

[2] National Fire Protection Association, 2001, *NFPA 1720, Standard for the Organization and Deployment of Fire Suppression Operations, Emergency Medical Operations, and Special Operations to the Public by Volunteer Fire Departments* Section 1.1.1.

7

Comprehensive Scheduling

One of the most challenging aspects of being a volunteer training officer is managing the training calendar and achieving departmental training goals within the narrow frame of the volunteers' available time. Not only must the training officer find space on the calendar for recruit, basic, and advanced training, they must avoid the major holidays, vacation seasons, and sporting events if they hope to have acceptable attendance. Bringing all of the various components of the training system to one calendar is not easy and requires an individual to be comfortable with multi-tasking to survive in the long term. This chapter will examine how to balance the various training system components with the limited amount of time available for the volunteer firefighter to commit to training.

Scheduling for Family

To gain support of today's demanding training schedule, the training officer must understand how to maximize the time available to today's volunteer. To do so, the officer must understand the demands and concerns of the modern family. The average volunteer is between the ages of 30–39 years, a white male, with a high school education. Most are married, blue-collar workers who statistically serve less than five years.[1] This five-year half-life for volunteers is an alarming statistic. Considering the costs involved outfitting and training a volunteer, getting only a five-year return is barely breaking even. No sooner has the volunteer training officer trained the recruit to be a self-sufficient, fully functional firefighter, than they quit the department. Finding ways to increase the five-year ceiling is the underlying mission for both the volunteer training officer and recruitment committees.

To find success in meeting the department training mission, the training officer must understand the various factors that pull at the volunteer for their attention. Their families, jobs, and hobbies are all competing for their time and trying to wedge into the space. The extensive training required of today's fire service is nearly impossible. The volunteer training officer must realize the priorities of life that should remain unchanged and make them the credo of the training program.

Family comes first! The training officer must be the first to admit that no matter how noble the fire service, the family must come first. Just hearing the training officer say these words can ease the minds of the volunteer firefighters. It should be policy that those special family events, such as that championship little league soccer game, take precedence over fire training. The training officer should be prepared to assist volunteers who must attend these events by making up the work. This is not to say the training officer must accommodate every request. Those firefighters who are proactive and notify the training officer of scheduling conflicts ahead of time receive the makeup; those who fail to communicate such needs and simply ignore the training requirements are on their own to make up the work. Experience has shown that in many cases, volunteers who truly have family conflicts can make up the work ahead of class, enabling them to keep up with the class easily.

A common oversight made during recruitment efforts is ignoring the potential volunteer's family. Many departments recruit mom or dad for service and never meet the families involved. This sometimes places the volunteer in a position where they are torn between family and duty, and in most cases, family will prevail, as it should. Through training, the family as a whole can be introduced to the volunteer system and gain valuable understanding about the contribution they are making to their community. Without this exposure to the department, the family is more likely to become resentful of the time spent away from home and the tug of war will begin. When building the training calendar, be sure to include time for the family to witness or be a part of the training operation. Remembering that recruitment and training involves the family unit as a whole will enhance retention efforts and make training a family goal, rather than having a negative impact on the family unit.

Managing the Training Calendar

Creating a central training calendar is essential to managing the various classes and tasks assigned to the training officer. The training calendar serves as the skeleton for the training system, providing a formal structure onto which the various components hang. The calendar should be accessible to both the instructor staff and students so that the information is available to anyone interested. Advanced planning using a training calendar will help the training officer manage the overwhelming schedule by ensuring plenty of time to prepare each class.

There are two main formats for training calendars, the traditional paper calendar and the electronic calendar. Departments with limited resources may choose to simply buy a printed calendar and post it in the fire station. This works well as long as the calendar is accessible to everyone as the central location for getting information about training. Departments blessed with more resources may have access to electronic calendars as part of a software package. This allows the courses to be entered as appointments and makes it easy for the training officer to organize training. The more difficult aspect of the electronic format is that it must be accessible to all members. This can be as easy as periodically printing an updated calendar and posting it in the station, or as technical as posting the calendar on the Internet or large area network (LAN). Whatever the situation, make sure that all members can access the calendar at their convenience.

Some simple components should be present in any format of training calendar to ensure that the appropriate information is available to all. Each entry should include the start time of the course. Establishing a standard weeknight and weekend start time will help to ease confusion for departments that run a high volume of training. For example, the training officer can establish that all evening classes will start at 18:30 and all weekend classes start at 08:00. This provides the system some rare predictability and cuts down on some of the confusion around the various classes. The entry should also include the end time. This enables the volunteer to plan around the training because they can schedule things after the class, and gives the instructor a limit on how long the training will last. Open-ended time frames can lead to run-on sessions that quickly turn boring and invaluable to the volunteer.

The entry should include the location in which the training will be held. For small departments, this may simply mean noting Training Room with the entry, whereas larger departments may indicate a classroom number. Be sure to include all of the spaces that will be required for the course. This will prompt the training officer to book additional resources when necessary and provides a means for the areas to be spoken for to prevent others from intruding on the class. For example, a session with lecture and skills training may require the training room and the apparatus bay. Indicating such will prevent the local girl scouts or other organization from double-booking the space.

Along with noting the location, the training officer should list any tools or equipment that may be needed for the program. This information can be entered in the notes section of the calendar software, or in small print on the paper calendar. This entry will help everyone involved plan what will be required for the training and again prompts the training officer to do advanced planning. This procedure will also prevent the double-booking of fire apparatus or other specialized equipment. All parties should know to check the training calendar to see if a unit or piece of equipment is available.

Planning the training calendar can be done on a yearly, quarterly, or monthly basis. The preferred method of many training officers is to create an annual calendar to put the various system components in place. Annual calendars make it easier to spot holes in the schedule or periods where the calendar may get too busy. The training cycle should reflect the department's budget cycle. This will make things less confusing by providing a defined starting and stopping point for the various classes and makes creating annual reports to support budget requests easier and more effective.

The first step in developing a training calendar is to do the homework to identify any holiday or special events that would interfere with the training program. This includes those little known holidays that many forget about and special department events, such as major fundraisers and socials. These occurrences should be blocked out in the calendar to ensure that no classes are scheduled in conflict with other events. Care should be taken to not schedule too much training during major holiday seasons such as Christmas and Thanksgiving because many families travel and may need a few extra days off. Scheduling light during the summer months will also ease some of the tension for the volunteer firefighter due to the family vacation season. Once all of the major events are boxed out, the training officer can see exactly which days are available on the calendar for training.

It is also advisable to build buffers into the training calendar. Build in time to absorb class cancellations due to adverse weather conditions or major incidents that require training to be cancelled. This may mean leaving a few extra nights or weekends open for emergencies. For academy programs, these buffers can be marketed as spring or fall breaks, giving students the feeling of having a vacation. By placing a week off in the middle of the longer programs, the training officer will have time to reschedule training should a major event cancel classes.

There are two schools of thought as to how to schedule individual programs. Some advocate using a consistent pattern to lay out the training dates; for example, using a Monday–Wednesday–Saturday schedule consistently throughout a program. Others advocate mixing the schedule up, using a Monday–Wednesday–Saturday rotation one week, and a Tuesday–Thursday–Sunday schedule the next. This second method is intended to provide those who have other commitments, such as softball teams or school, a chance to at least make half of the training, and make up the rest. A compromise in this approach is to run a particular program in Monday–Wednesday–Saturday for one full presentation of the course, and the next time the course is offered, use the opposite schedule. This works well as long as the course comes up again in the rotation at a regular pace.

Whether using a simple paper calendar or a complex computerized system, the training officer must ensure that the training calendar is maintained in an accurate and timely fashion. Ensuring that all members have access to the calendar will help the training officer in their mission of training the department.

Incorporating Incentives

An alternative or supplement to simply requiring volunteers to attend training is to incorporate incentives into the program. At first glance, training officers may overlook this option because of the associated costs in providing incentives. A closer looks identifies incentive options that are relatively cheap and require little effort to implement. Available resources will dictate how elaborate any incentives programs may be.

Most programs award prizes based on either simple attendance, offering awards to the volunteer who obtains the most training hours, or, on performance standards for the volunteer with the best scores. The training officer must determine the best method to award incentives that is consistent and stands the test of time. A record-keeping method must be established that provides a means for volunteers to qualify for the program.

Getting the Calendar Started

Using one of your own choosing, identify the beginning of the department budget cycle and begin planning with that month. Mark off any holidays or special events that would impact attendance for training. What is left may be used in the planning process. As you work through the next sections, start to schedule training in layers, applying the most basic courses in the mission first, then adding other training goals for the year.

Basic firefighter rotations

Before scheduling basic firefighter rotations, the training officer must determine how they will handle the transition of new volunteers to active members riding on the fire apparatus. The training required to ride the fire apparatus to emergencies must be identified and should be at a minimum, enough to ensure recruit safety for exterior operations on a working fire. The fireground operations course discussed in chapter 4, and offered on the CD-ROM, is recommended as a minimum volunteer recruit standard.

There are two ways to approach recruit firefighter training. Some may choose to run academy level training, or complete firefighter certification programs frequent enough that a supplemental orientation program may not be necessary. This, however, is a luxury reserved for large departments with enough resources to start another academy level class every two or three months. Most volunteer departments must use an orientation program as a band-aid to keep new volunteers active so they will not lose interest until the next academy level class begins. Because this is a common practice in the volunteer setting, the following section will address both volunteer orientation training and academy level training. Larger departments that have sufficient resources may choose to have volunteer recruits go directly to the academy level because their turnaround time between the start of the next available class is limited.

Volunteer orientation training

The key to scheduling orientation training is making it frequent enough to keep up with the influx of new volunteer recruits, yet kept at a pace that the training officer can withstand and maintain their sanity. A six-week rotation is recommended for departments that recruit 50–100 new volunteers annually. Departments that recruit 25–50 new volunteers annually may expand the rotation to as long as two to three months, depending on their specific needs. The schedule may also be adjusted to reflect any annual recruitment drives that may occur, placing more training in the busy season and slowing the pace during other times. Departments that recruit fewer than 25 new volunteers annually may wish to provide the training on a one-on-one basis, assigning an instructor to work directly with one or two new volunteers, instead of having a formal course schedule. Remember that the orientation training is performance-based and students need not attend every class session, but they must demonstrate their ability to meet the course objectives.

Using a calendar starting with the beginning of the department's budget cycle, add the 24-hour volunteer recruit training as described in chapter 4. The schedule provided with the program is based on a Monday–Thursday schedule, two weeks in a row. This may not be appropriate for all departments and training officers may wish to alter the schedule to meet their specific needs. Once the first two-week session is established, simply count off the weeks until the next session should start based on the amount of recruitment for the department. For example, departments recruiting 50–100 volunteers annually should count four weeks from the end of the last course to find the beginning of the next. This will provide a one-month break between courses for the training officer. Lay out the full volunteer orientation for the year on the calendar, look for any conflicts with holidays or other events, and shift the schedule accordingly. Remember to indicate the location scheduled for the training also to prevent double-booking and to ensure your training space.

The more regular the schedule, the more predictable it will be, allowing fire officers to figure the start dates on their own. By addressing the most basic component of the training program first, the training officer is assured that this mission will be met with appropriate time and resources.

Academy or certification training

The academy program requires a great deal of planning to ensure that all of the course objectives are met, while accounting for what the volunteer can bear in the schedules. In determining the frequency of the program, the training officer should return to the recruitment statistics to determine what will best suit their department. The academy program works best with no more than 30 students at a time. Departments that recruit more than 50 volunteers annually should expect to run two programs a year. This places a six-month window on the program for completion without overlapping. Departments with recruitments of less than 30 per year are not confined to the six-month schedule and may spread their academy schedule over the six-month deadline.

Some training officers may be faced with training their annual volunteer recruitment at the academy level, plus some members of their department who, for whatever reason, have failed to obtain the minimum training requirements. In this case, the training officer must determine the total number of academies required for their recruitment, plus provide a means for addressing the department backlog. Typically, a long-range plan is required to address such issues, as most training officers find their systems running nearly at capacity simply training new recruits. Training officers may develop plans to train their backlog a few members at a time until the job is finished. A large backlog may require a commitment from the entire department to support another academy in the schedule, either by the commitment of funding or staffing. This may require training officers to run multiple academies simultaneously or overlapping. In this situation, training officers must fit their academy schedules into a six-month timeframe to enable some sort of time limit and allow for some basic efficiency. Training officers not facing this challenge may relax their schedules over the six-month limit if only one academy is required annually.

When working within the six-month time frame, the best approach for scheduling is a two-night-per-week, every-other-Saturday schedule. This pace is about all the training the volunteer can stand and still juggle responsibilities at home and at work. Getting all of the topics into a six-month schedule and maintaining an every-other-Saturday rotation requires some creative scheduling and causes some weekends to be doubled up toward the end. With this scheduling method, the brunt of the scheduling burden is saved until near the end of the program when the student can see the light at the end of the tunnel and the training is practical, live fire training; something most students seem to enjoy. The students will spend an occasional Saturday/Sunday in the burn building, but at least that's better than 16 hours in the classroom.

Training officers forced to use overlapping or simultaneous academies can opt for a Monday-Wednesday-Saturday rotation running in conjunction with a Tuesday-Thursday, the other Saturday schedule. This also allows students who miss a Monday night class to make up a session on Tuesday night, etc. This type of scheduling is brutal on the training staff, but may be necessary to overcome training deficits.

Working from the Academy Syllabus Template (Fig. 7–1 and also available in the included cd-rom), work with the dates on the calendar using the two nights per week formula. Courses that list an 18:30–22:30 time slot are four-hour blocks best suited for evening sessions. This is not to say that two evening sessions can't be doubled up to make a Saturday session. The training officer must work with the schedule to find the best way to work through holidays and events. The 08:00–16:30 sessions are intended to be Saturday sessions with a half-hour lunch break. Figure 7–2 (also available in the cd-rom) shows an actual finished course syllabus that utilizes this scheduling theory. Notice that reading assignments are made well in advance to allow the student plenty of time to prepare. Once the academy schedule is established, the dates can be entered in the calendar along with the previous orientation training. *Using your training calendar, add the proper number of academies to the schedule based on your recruitment statistics.*

Academy Syllabus Template

Date	Time	Topic	Reading Assignment	Equipment Needed
	18:30 — 22:30	Orientation (FF1) (FF2)		
	18:30 — 22:30	Safety (FF1) (FF2)		
	08:00 — 16:30	Drill and Ceremony/Physical Assessment (FF1) (FF2)		
	18:30 — 22:30	Communications (FF1) (FF2)		
	18:30 — 22:30	Fire Prevention (FF2)		
	18:30 — 22:30	Public Fire Education (FF1)		
	18:30 — 22:30	Fire Detection and Systems (FF2)		
	18:30 — 22:20	Fire Cause and Determination (FF2)		
	18:30 — 22:30	Sprinklers (FF1)		
	18:30 — 22:30	Fire Behavior (FF1)		
	18:30 — 22:30	Building Construction (FF1) (FF2)		
	18:30 — 22:30	PPE Lecture (FF1)		
	18:30 — 22:30	SCBA Lecture (FF1)		
	08:00 — 16:30	PPE/SCBA Practical (FF1) (FF2)		
	18:30 — 22:30	Fire Hose Lecture (FF1) (FF2)		
	08:00 — 16:30	Fire Hose Practical (FF1)		
	08:00 — 16:30	Fire Streams (FF1) (FF2)		
	08:00 — 16:30	Fire Control (FF1) (FF2)		
	18:30 — 16:30	Ground Cover Fires (FF1)		
	08:00 — 16:30	Water Supply (FF1) (FF2)		
	18:30 — 22:30	Vehicle Fires (FF1)		
	18:30 — 22:30	Fire Extinguishers (FF1)		
	08:00 — 16:30	Rescue and Extrication (FF2)		
	08:00 — 16:30	Rescue and Extrication (FF2)		
	08:00 — 16:30	Search and Rescue (FF1)		
	08:00 — 16:30	Ropes and Knots (FF1)		
	08:00 — 16:30	Forcible Entry (FF1)		
	08:00 — 16:30	Ladders (FF1)		
	18:30 — 22:30	Ventilation (FF1)		
	08:00 — 16:30	Salvage and Overhaul (FF1)		
	18:30 — 22:30	Basic First Aid (FF1)		
	08:00 — 16:30	CPR (FF1)		
	32 Hours	Hazardous Materials Operations		
	18:30 — 22:30	Night Evolutions (FF2)		
	18:30 — 22:30	Night Evolutions (FF2)		
	08:00 — 16:30	Mayday! Firefighter Down (FF2)		
	08:00 — 16:30	Mayday! Firefighter Down (FF2)		
	18:30 — 22:30	Report Presentations (FF2)		
	08:00 — 16:30	Burn Evolutions (FF2)		
	18:30 — 22:30	Skills Testing (FF1)		
	08:00 — 16:30	Burn Evolutions (FF2)		
	18:30 — 22:30	Night Evolutions (FF2)		
	08:00 — 16:30	Burn Evolutions (FF2) Family Day		

Fig. 7–1 Academy Syllabus Template

HANOVER COUNTY FIRE DEPARTMENT
Training & Quality Assurance Division
Hanover Fire Academy
P. O. Box 470
Hanover, VA 23069-0470

Fire Academy
Class of 2002 — School #09
(Tuesdays, Thursdays & Every Other Saturday)

Firefighting Training

Date	Topic	Time	Preparatory Assignment
07/09/02	Orientation	18:30	Read Chapter 1 (5-20)
07/11/02	Safety	18:30	Read Chapter 1 (20-30)
07/13/02	Drill & Ceremony/Physical Assessment	08:00	
07/16/02	Com munications	18:30	Read Chapter 18
07/18/02	Fire Prevention	18:30	Read Chapter 19
07/23/02	Public Fire Education	18:30	
07/25/02	Fire Alarms & Communications	18:30	Read Chapter 15 (559-571)
07/27/02	Cause & Origin	08:00	Read Chapter 17
07/27/02	Fire Sprinklers	13:00	Read Chapter 15 (571-583)
07/30/02	Fire Behavior	18:30	Read Chapter 2
08/01/02	Building Construction	18:30	Read Chapter 3
08/06/02	Personal Protective Equipment	18:30	Read Chapter 4 (79-86)
08/08/02	Self Contained Breathing Apparatus	18:30	Read Chapter 4 (86-111)
08/10/02	SCBA Practical / **Skills Testing**	08:00	
08/13/02	Search & Rescue	18:30	Read Chapter 7 (175-186)
08/15/02	Search & Rescue Cont.	18:30	
08/20/02	Ventilation	18:30	Read Chapter 10
08/22/02	Fire Hose	18:30	Read Chapter 12
08/24/02	Fire Hose/**Skills Testing**	08:00	
08/27/02	Forcible Entry	18:30	Read Chapter 8
08/29/02	Forcible Entry Cont.	18:30	

Fall Break

Date	Topic	Time	Preparatory Assignment
09/10/02	Ground Cover Fires	18:30	
09/12/02	Vehicle Fires	18:30	
09/14/02	Fire Streams	08:00	Read Chapter 13
09/17/02	Portable Fire Extinguishers	18:30	Read Chapter 5
09/19/02	First Aid / **Skills Testing**	18:30	
09/24/02	Ropes & Knots	18:30	Read Chapter 6
09/26/02	Ropes & Knots Cont.	18:30	
09/28/02	Fire Control / **Skills Testing**	08:00	Read Chapter 14
10/01/02	Water Supply	18:30	Read Chapter 11
10/03/02	Water Supply Cont.	18:30	
10/08/02	CPR	18:30	
10/10/02	CPR Cont.	18:30	
10/12/02	Vehicle Extrication	08:00	Read Chapter 7 (186-215)
10/13/02	Vehicle Extrication	08:00	
10/15/02	Salvage & Overhaul	18:30	Read Chapter 16
10/17/02	Salvage & Overhaul Cont.	18:30	
10/22/02	Haz-Mat Operations	18:30	Read Student Guide
10/24/02	Haz-Mat Operations	18:30	
10/26/02	Ground Ladders	08:00	Read Chapter 9
10/29/02	Haz-Mat Operations	18:30	
11/05/02	Haz-Mat Operations	18:30	
11/07/02	Haz-Mat Operations	18:30	
11/09/02	Haz-Mat Operations	18:30	
11/12/02	Haz-Mat Operations	18:30	
11/14/02	**Live Fire Evolutions**	18:30	
11/16/02	**Live Fire Evolutions**	08:00	
11/17/02	**Live Fire Evolutions**	08:00	
11/19/02	**Live Fire Evolutions**	18:30	
11/21/02	LODD Report Presentations	18:30	
11/23/02	Mayday–Firefighter Down	08:00	
11/24/02	Mayday–Firefighter Down	08:00	
12/03/02	Academy Skills Testing	18:30	
12/05/02	**Live Fire Evolutions**	18:30	
12/07/02	**Live Fire Evolutions**	08:00	
12/10/02	State Written Exam (FF1)	19:00	
12/12/02	State Practical Exam (FF1)	19:00	
12/14/02	**Live Fire Evolutions (Family Day)**	08:00	
12/19/02	State Written Exam (FF2)	19:00	
01/09/03	Academy Graduation	19:00	

Fig. 7–2 Sample Academy Student Syllabus

Once the basic mission of recruit and academy level training has been met, the training office can evaluate what remains open on the calendar and prioritize what else needs to be accomplished with the days remaining. Most will find few dates left, especially departments required to run two or more academies in a year.

Specialty and advanced scheduling

Specialty and advanced training topics include any topic beyond the basic Firefighter I and II levels. This includes fire officer training, driver training, and anything else that may be required throughout the year. The training officer must first determine what is required by the department, such as training requirements for elected or appointed positions, and ensure that such classes are made available at least annually. Make a list of any such training, including the total hours required for each. Many of these courses fall in the 16- to 32-hour range and provide the training officer a variety of scheduling options. The courses may be conducted in full 8-hour blocks on a Saturday and Sunday, or they may be broken up into 4-hour blocks and spread out over a month. The training officer must determine the best course of action based on what is available on their calendar. Working from the list of required training, examine the schedule for areas that seem light. Work with the calendar to find the best way to enter each required subject until all required topics are entered.

At this point, the training officer should have met the basic training mission for the department, including volunteer recruit and basic training, plus the other required topics indicated through department bylaws or training requirements. The training officer can now start to consider training from the *wish list*, be that from their own ideas of what training should be done, or that of their peers. Many officers will find the calendar chocked full of training, leaving almost no days open. Adding the classes from such wish lists can seem impossible from both the available time and space, and from instructor staff and resources points of view. Applying the courses on the calendar in this manner gives the training a firm platform on which to stand to ask for the resources necessary to go beyond the basics. Many volunteers, including officers, fail to realize the impact that each class added to the schedule has on the overall training system. By showing the system in calendar form, the training officer can better make a case for increased funding and resources. This will help prevent internal unfunded mandates that can quickly overrun the training system.

In-service scheduling

The only thing more difficult than choosing in-service topics is finding the time to actually conduct the training. Even though one of the key factors in successful in-service programs is finding new and exciting topics, the key to scheduling such training is making it as predictable and routine as possible. This enables volunteers to work around the regularly scheduled training as best they can. Many training officers choose a regular night of the month for in-service training, such as every third Monday. In this case, the volunteer knows months in advance that the third Monday is training night and this gives them a chance to plan well in advance to attend.

For larger volunteer departments, just one training night will not meet the demands of the department. Multiple training nights may be required to accommodate the sheer numbers of volunteers who must attend. This may require repeating the training over a three-night period, for example, or offering the training to each station, should the training officer have multiple companies in their organization.

The Hanover Fire Department uses a multi-phase approach to in-service scheduling. Having 12 companies within the department, each company has a training officer who offers in-service training based on individual company needs, typically once a month. The county training officer schedules in-service training monthly, but at a different company each time. This enables the company training officer to rest for a month if they wish or to have two separate training sessions in a month. The county in-service classes are announced countywide, which allows volunteers from all over the county the option of attending.

The training officer must remember to incorporate the company officer as an assistant in implementing in-service training and ensure that they are active in the instruction process. By scheduling the training so that it is predictable yet interesting to the volunteer, they will find the challenge of in-service training a fun and interesting part of their job duties.

Developing Scheduling Templates and Formulas

Once the training officer has completed the first year's training calendar using a systems approach, they have developed a template to use in the future. The training officer can simply monitor recruitment statistics to determine their academy needs and make adjustments for the other topics accordingly. A formula is developed that allows the training officer to quickly plan their year and begin advanced planning to obtain instructors and resources. Remember, this scheduling formula should detail courses identified through a needs assessment and must be based on recruitment and retention data. Figure 7–3 (also available in the cd-rom) provides a sample scheduling formula used by the Hanover Fire Academy. It includes the basic courses required each year and allows the training officer to quickly create a training calendar.

Category	Offered Annually	Topic	Hours	Identified Via
Recruit	10 (6-week rotation)	Fireground Operations Course (orientation)	24 (8-night sessions)	Recruitment statistics
Basic	2 (6-month rotation)	Volunteer Fire Academy • Firefighter I • Firefighter II • Haz-mat operations • Mayday-FF Down • Vehicle extrication • CPR	232 (2 nights/ 1 Saturday)	Recruitment statistics
Advanced/specialty	1	Incident Command Systems	16	SOP
Advanced/specialty	1	Leadership I1	6	SOP
Advanced/specialty	1	Basic Pump Operator	16	SOP
Advanced/specialty	2	Rural Water Supply	16	Needs assessment
Advanced/specialty	3	EVOC	16	SOP
Advanced/specialty	Monthly	Vehicle Extrication —Awareness Level*	16	Needs assessment

*Vehicle Extrication Awareness is the startup phase of an Awareness-Operations-Technician Level retraining for the whole department. The course is offered monthly over a 3-year period to offer training to the entire response system.

Fig. 7–3 This Training Formula Offers a Systematic Approach to Scheduling that Accounts for All Parts of the Training System (SOP—standard operating procedure)

Sub-templates can be used to apply the same theory to individual courses that are complex, such as the academy program. Building a course template will aid the training officer in quickly scheduling courses that repeat each year, based on a simple scheduling format. Figure 7–4 (also available in the cd-rom) provides a generic scheduling template used by the Virginia Department of Fire Programs. It contains the basic information that will be required when scheduling most courses. By filling in this basic information, the training officer will have everything needed for marketing the program and for appropriate documentation of the training. It is important to restate that the templates provide only suggestions regarding scheduling. The training officer must be creative in finding ways to fit the required training into schedules that will meet their volunteers' needs.

Virginia Department of Fire Programs
Course Schedule Form

Course Name: _____   Course Number: _____

Date of Training	Day of Week	Subject	Type of Class (Classroom/Skill)	Time (Start/End)	Total Hours	Instructors

Fig. 7–4 Generic Scheduling Template (Courtesy Virginia Department of Fire Programs)

Notice that in Figure 7–1 (Sample Academy Template), the template calls for Water Supply to be held during an 8-hour Saturday session, yet in Figure 7–2 (Sample Academy Student Syllabus), the course is actually scheduled over two weeknights. This is done to make the scheduling work within the confines of the six-month period and to give the volunteer students the every-other-weekend rotation that is so vital to their success. Use the templates as a guide, not the final word, to get the schedule formed in a manner that meets the department's individual needs.

Chapter Review Questions

1. What is one of the most common oversights made by volunteer recruitment and retention programs?

2. What is the first step when laying out the training calendar?

3. What is a valuable tool when scheduling courses that repeat on a regular basis?

4. What is the recommended scheduling format for academy level programs?

Answers

1. Recruitment and retention programs often overlook recruiting and retaining the family in conjunction with the actual volunteer.

2. Blocking out any holidays or important events that may interfere with training.

3. Scheduling templates.

4. Two nights a week and every other Saturday.

Notes

[1] National Fire Protection Association, 2000, *NFPA 1410, Standard on Training for Initial Emergency Scene Operations* 1-1.2.

8

Alternative Training Options
and Marketing

E ven with a systems approach to volunteer training, the training officer may find a need to go outside the box to meet the various training needs of the department. This requires a mindset that allows goals to be met using imaginative methods that may require a combination of resources with neighboring departments or finding funding in unlikely places. It is also important that the training officer be able to market their training system in such a way as to maximize the efficiency of the system. Even the best training programs will fail if no one is aware that the program exists. This chapter will examine some training alternatives and marketing methods that will help to ensure success of the volunteer training program.

Special Events Planning

Once the volunteer firefighter has completed their basic training requirements, a large part of their training needs to center around short 16- to 32-hour programs. These include leadership and incident management training, as well as specialty topics such as advanced rescue or pump operator courses. This type of training can have a wide range of logistics and support necessary for each and can become taxing on the volunteer training system. As opposed to addressing each course in a singular fashion, the training officer may find benefit in presenting the training packaged together as part of a special event. This enables the training officer to gain maximum efficiency for the logistical support required for such training and combining the courses provides a unique marketing tool that can help increase attendance.

As usual, the need for a special event format is identified through a needs assessment. Upon recognizing a need for a variety of short-term training classes (less than a couple of weekends per class), the training officer can identify which programs would be appropriately packaged together and determine the scope of the event. This will help identify the venue requirements by estimating how many students may attend the event and other logistical needs. Figure 8–1 offers a sample list from a past special event that provides some examples of such training. For the event, instructors were brought in from around the country to present a fresh perspective on the basics and were teamed with local instructors to provide courses identified through a needs assessment. With more than 300 people in attendance, the training event proved to be a good investment at a cost of approximately $33 per student. Logistically, the benefit of this event was that a small training staff was able to offer a vast amount of training in one weekend by enlisting the help of contract instructors and a few friends to meet the training need. By putting all of the courses either under one roof, or in close proximity to each other, a limited staff can facilitate training that in any other setting would be overwhelming.

Course Title	Instructors	Hours
Chief's Seminar Preparing Fire Officers for Administrative Responsibility	Dr. David Hoover, University of Akron	16
Back To Basics Seminar Engine and Truck Company Operations	Rick Lasky, Chief, Lewisville, Texas Don Hayde, Battalion Chief, FDNY Sal Marchase, Lieutenant, FDNY Seth Dale, Darien-Woodridge Fire District, Illinois Tom Shervino, Captain, Oak Lawn Fire Dept. Illinois Curtis Birt, Darien-Woodridge Fire District, Illinois John Hojek, Jr., Lieutenant, Oak Lawn Fire Dept., Illinois Rich Collins, Division Chief, Darien-Woodridge Fire District, Illinois Mike Orrico, Oak Lawn Fire Dept, Illinois	16
Mayday Firefighter Down!	Virginia adjunct instructors	16
Confined Space Rescue	Virginia adjunct instructors	16
Vehicle Rescue	Virginia adjunct instructors	16
Incident Command Systems	Virginia adjunct instructors	16
SCBA Maze (Fairfax, VA)	Virginia adjunct instructors	16
EMT-A Recertification	Virginia adjunct instructors	24

Fig. 8–1 Example of a Regional School Using Instructors from All Over the Country to Give Volunteers a Broad Cross-Section of Courses to Choose

Many considerations must be addressed when planning such an event. Everything from providing refreshments and coffee to portable toilets must be considered to ensure that everything works smoothly. The training officer should create a workbook that outlines the specific information for each class for easy reference and document each plan made in preparation for the training event. The workbook should contain every detail, so that even if the training officer were to be incapacitated in some way, another officer could pick up the book and manage the event easily. Managing the project in this manner will help cut down on *he said/she said* controversies and may help protect the training officer should something go wrong. Making copies of the information just before the event and issuing it to the appropriate assistants will also provide redundant protection against any mishaps that are bound to come up.

One of the highlights associated with the event presented in Figure 8–1 was an opening ceremony that served several purposes. First, the ceremony enabled all of the students to come together in one auditorium and realize the scope of the training event. Seeing more than 300 people gathered on a Saturday morning is impressive by anyone's standards. Secondly, the ceremony provided coordination of the students and addressed the business side of the event, such as lunch and break schedules and restroom locations. The event also gave local politicians a chance to address the group and thank them for their commitment to emergency services. A ceremony also gets the volunteer fire department into the public eye and acknowledges the department's commitment to the community. The ceremony included bagpipers and a special ceremony honoring firefighters who have fallen in the line of duty and dedicating the training in their memory. This short 20-minute ceremony is well worth the effort and makes an excellent impression on firefighters from other jurisdictions.

Another feature was a seafood feast scheduled for Saturday night, after the first day of training. The fire service loves to eat and this provided an excellent chance for firefighters to network and discuss the weekend's events. The meal was provided by the volunteer association and was a highlight, making this particular event known throughout the region. The dinner also got all of the 300 plus students back together to marvel at the sheer size of the event. It also gave local firefighters a chance to mingle with the instructors brought in from around the country to share ideas and information.

Funding considerations

The one drawback to using a special event format to deliver training is the costs associated with such an event. There are several options available to the training officer to find the necessary funding. Full funding is the easiest option by gaining funds either through the department budget, or from a state organization or institution responsible for fire training. This requires a formal request well in advance of the event to allow time for the wheels of bureaucracy to spin up approval and funding. Another option is charging admission or tuition to cover the costs. In Figure 8–1, the students were charged a $30 fee that covered the seafood feast, which was all-you-could-eat and included beverages. The actual cost of the training was derived through the Virginia Department of Fire Programs and was supplemented by the department budget. For either the training itself or the seafood feast, the cost per student was approximately $30.

To determine a potential tuition cost for a special event, the training officer must calculate the total cost of the program, including everything from instructors, to teaching supplies, to refreshments. Divide the total figure by the minimum number of students required to have the school. Using the minimum attendance requirement rather than the maximum will ensure that the project doesn't generate a deficit and may even generate a funding surplus that can be used to address those emergencies that are bound to arise.

Seeking out corporate sponsors is an effective way to ease the cost of the training on both the department and the students. Having such a large number of firefighters in one place is a dream come true for many fire service vendors and they might be willing to provide funding in exchange for getting their name and/or products out to the group. This may be done through full or partial sponsorships that can generate enough funding to meet the training needs without charging ridiculous tuitions for the event. The training officer may wish to go outside the fire service industry to find community sponsors willing to help in the cause. Whether funded fully by the department, through another agency, through sponsorships, or a combination of all three, the training officer can muster the resources for such events in almost any situation. It requires advanced planning and an ongoing needs assessment to identify funding sources early.

Locating facilities

In most cases, the volunteer training officer will find they lack the facilities necessary to host several hundred people at once. This requires the training officer to look throughout the community to identify potential locations for events as they come up. The local school system provides an excellent resource, as most provide access to auditoriums, gymnasiums, and classrooms necessary to produce most training events. High schools are preferred over lower schools, as the facilities and seating are geared more towards adult learners. The seating in lower schools will be geared toward smaller students and will likely be uncomfortable for some fire-fighters. When high schools are unavailable, the training officer may consider nearby colleges as an alternative.

Establishing an ongoing relationship with the appropriate school officials will help the training officer when the facility needs arise, whether it be a sudden situation or an advanced plan. Learn the process for obtaining such facilities and follow them to the letter. The training officer must realize that they are competing with other school functions for space, and when a conflict arises, the school will come out the winner. To combat such situations, the training officer must reserve the facilities as soon as they confirm a date and time. The earlier the event is on the school's schedule, the less likely the training event will be bumped, by say, the drama club. Remember that schools run on a seasonal schedule and the training officer will be dodging football, prom, and graduation seasons.

When schools are unavailable, the training officer may consider local churches, hospitals, or any other location that may have classroom facilities. Most are happy to partner with the local volunteer fire department to protect the community. The risk in using such locations is that in most cases, the firefighters may share space with the owners of the facility. Getting more than 10 firefighters in the same room runs a risk of offending almost anyone (much less 300 firefighters), and the training officer must have measures in place to protect the relationship with the venue. It is also vital that the training officer ensure that the facility is left in appropriate condition. It is likely that the training officer may have to call on that facility again in the future and facilities are unlikely to allow organizations to use their space when it is not done in a respectful manner.

Managing logistics

A variety of unpredictable needs may challenge the training officer during a special event. No matter how well planned the event is, something will go wrong or be overlooked. To accommodate such occurrences, the training officer must ensure that a response is standing by for almost any occasion. Assigning a team of individuals with rapid access to resources such as water and ice, food, audio-visual (AV) equipment, etc. will help address those sudden emergencies that happen from time to time.

The best management approach to such events is to treat it like any other emergency. Assign responsibilities for various sectors and divisions to provide adequate management at the various locations. Figure 8–2 provides a sample organization structure for managing the special events courses identified in Figure 8–1. The chart uses standard incident command theory and simply changes the names to match the job description for each position. Most of the positions are self-explanatory, but some may require explanation. The instructor coordinator is responsible for making sure that the various instructors show up on time and have anything that they might need. These needs can be communicated through the logistics officer to the AV task force that is designed to have at least one of everything related to audio-visuals. This includes extension cords, projectors, duct tape, etc. Having a school representative available, at least by radio, enables the training officer to quickly tap extra resources or make minor adjustments in room temperature or access to restrooms. The mechanic should be on standby via radio contact in case a unit breaks down. Using standard incident management techniques, the training officer can manage a large-scale event in an effective manner.

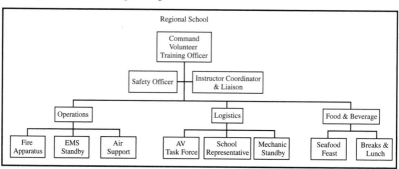

Fig. 8–2 Special Events are Run Much Like Emergencies (an organizational chart can be created to coordinate the operation)

Demobilization

A common oversight by training officers when dealing with special events is wrapping up the event. Massive amounts of planning are employed right up until the program ends, leaving the termination of the event an unorganized mess and leaving a bad last impression on all involved. A plan must be developed to systematically demobilize the event right up until the doors are locked. Huge amounts of resources are deployed rapidly during the course of the school, and it is easy to overlook items in the rush to go home.

The training officer begins the demobilization process about midway through the last day. Inventories must be done on all equipment, and care must be taken to ensure that the facility is left in *better* condition than it was found. As individual classes conclude, the training officer can dispatch crews to ensure that each room is left exactly as it was found and that all equipment is recovered and loaded for transport back to its original location. Using a digital or Polaroid camera to document the condition and placement of items in the room, is an excellent way to ensure a clean exit of the facility. That way, if any claims are made about the condition of the facility after the event, the training officer can refer to the photo to determine the facts.

Marketing Training

In the volunteer fire service, many valuable training programs come off as a failure, not because of the content of the program, but because it was poorly marketed. It is vital that the training officer understand the basics of marketing and how it affects their ability to provide effective training to their department. Understanding these concepts will help the training officer maximize their effectiveness and improve attendance statistics.

Most people think of marketing as it relates to television commercials and print ads common in the modern marketplace. Many volunteer training officers fail to realize the positive impact a conscious and effective marketing strategy can have on their training system. The following section will apply standard marketing tactics specifically to the volunteer fire service, enabling the training officer to get the most out of their training system.

In the business world, marketing consists of creating, promoting, and delivering goods and services to consumers and businesses.[1] This can easily be applied to fire service training as the training officer's tasks of creating training programs, announcing them to the membership, and actually delivering the training. Each part of the process plays a vital role in ensuring the success of the training effort. Knowing how to capitalize on each section gives the training officer the upper hand in ensuring that their training programs are successful.

The training officer must consider the marketing mix, which is a set of marketing tools used to pursue marketing objectives.[2] Marketing expert Jerome McCarthy identified the Four P's of Marketing as the components that make up the marketing mix used by businesses to determine the best course of action when considering new products or services (Fig. 8–3). The Four P's are product, price, promotion, and place. Each category provides key marketing points that the training officer can use to develop the best marketing mix for their program.

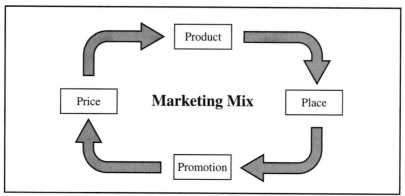

Fig. 8–3 Marketing Training Should be a Mix of Various Considerations

Product

Product describes the variety of training available within the training system. The various *products* must be designed to provide quality to the volunteer by maximizing their efforts in training. Training products offer something to the volunteer that they specifically need or want. In some cases, the product itself is enough to convince the volunteer to buy. The rush of firefighter survival training that took place in the late 1990s is an excellent example of the product selling itself. The creators of such training recognized a need in the fire service and created programs that sold simply on their content. Classes sold out repeatedly, not because of glossy advertisements, but because the content was good and in demand by the masses.

Packaging for the training program can have a huge impact in its success. Training officers must consider the best method to present their programs to the target audience. For each class, determine who needs the training and package the program to highlight the needs of the target group. For example, firefighters tend to enjoy hands-on training, especially training involving live fire evolutions. The training officer should ensure that the opportunity to get hands-on, live fire training is at the top of the training announcement and is the selling point of the program. Trying to sell a four-hour lecture on flashover is much more difficult to promote than a four-hour run in a live-fire flashover simulator. Determine what the target audience wants, and do your best to give it to them.

Some training systems can *sell* training based simply on the brand name. The National Fire Academy (NFA) serves as a good example of this theory. Firefighters from across the United States and abroad know that NFA classes will provide a minimum level of quality, simply based on the NFA logo being attached to the program. Every training officer can set this as a future goal. By striving for constant quality and improvement, training officers build a reputation that spreads throughout their area, facilitating improved attendance rates based solely on reputation.

Price

In most marketing theories, price refers to a monetary cost for purchasing a product. This theory can be applied to volunteer training, as some departments charge tuition for training. When using this practice, the training officer must use care to ensure that the cost associated with training meets their budgetary needs while not overwhelming the volunteer's personal income. Charging the volunteer to give their time should be a last resort and only used when the need for a particular class drives the tuition. In fact, the training officer should look at price as it relates to the training's impact on the volunteer.

To really understand the price of training, the volunteer training officer must think of the impact on the volunteer far beyond any tuitions or fees. Consider time away from the family, the cost of gas for driving to and from the training location, time spent preparing and studying, and cleaning up after class. The true price of training reaches far beyond the classroom. Be sure to factor in the full impact when thinking of price as it can have huge impacts on the training program.

The training officer can capitalize on the price of training by promoting a streamlined and efficient program. Simply recognizing how training impacts the volunteer can improve attendance, if they feel that they are understood. Volunteers will prefer any program that saves them time in the end. Advertising what the firefighter gets for the cost and ensuring that their time will be spent wisely will help the training officer better promote their programs.

Promotion

Promotion is the method used to communicate the appealing factors about a product to the public. In this case, it refers to how training programs are announced to the department. Many volunteer departments have a central location for training announcements, such as a training bulletin board or calendar. On the surface, this simple task of posting announcements seems simple, but upon closer examination, there is much more than initially meets the eye. The following section will examine key components of a training announcement and methods used to communicate training to the department.

Training announcements

Each training announcement should contain some standard information about the training program to help students make plans to attend and to cut down on questions. A template can be created that enables the training officer to rapidly create training announcements and ensure consistency from course to course.

The title serves as the heading for the announcement and must stand out from the rest of the page. Remember that training announcements are competing with other postings to catch the reader's eye on the bulletin board. The title should be related to the actual training topic in a way that sparks interest in the firefighter. The developers of Virginia's firefighter survival program struggled with naming the program in a way that would capture the essence of the course, yet be short and to the point. The title, "Mayday! Firefighter Down," was chosen because it was short and applicable, but more so, because it sparked interest. It did much more to promote the program than some of the other titles considered for the program, such as Advanced Firefighter Rescue Techniques. Based on the title alone, firefighters may perk to attention. Today the course is commonly referred to as the Mayday Program and even has its own line of tee shirts and golf-shirts that help promote the program.

The location of the training should be stated plainly on the training announcement. The location should be highlighted if the training officer plans to capitalize on the place factor of marketing by offering a new or unique training location. By plainly stating the location, including travel directions if appropriate, the number of questions and amount of confusion can be reduced.

The time of training should be stated, including start and end times. This enables the volunteer to make arrangements to attend, including arranging for babysitters and letting their spouses know what time to expect them home. This is extremely important for those who rely on others for transportation, as they may need to be picked up after the conclusion of the training session.

A written description of the training can give the training officer a chance to provide a *teaser* to the firefighters as to what they might expect from the course. The description should be truthful, but highlight the components of the program that capitalize on the marketing mix, targeting volunteers based on product, price, and place. This is an opportunity to sell the training to get the maximum attendance. Ensure that the description inspires interest in the target audience.

Prerequisites are a common omission on training announcements, yet they provide a tremendous benefit when posted appropriately. This section of the announcement should describe any prior training required to attend the upcoming course. It may also include any other special requirements, such as personal protective clothing or SCBA. Identifying the requirements in advance may prevent firefighters from making an effort to attend training that they aren't ready for and it may also prevent any harsh experiences that could take years to wipe away.

Methods of communication

A common complaint about training is that the potential student didn't know about the class or didn't find out about the training in time to prepare to attend. This inability to communicate upcoming training events can cause classes to be conducted with empty seats and can even create animosity toward the training officer. This leads to reductions in efficiency and drives up the cost per student ratios. Understanding how to effectively communicate upcoming training will help the training officer get the most return on their efforts and create a positive impression on the department.

Print, Internet, and verbal communications make up the basic options for the training officer to communicate training. Each can play a vital role in training communications and should be used whenever possible to ensure that all members are kept updated. These communication methods can be used in a singular fashion or in conjunction with one another to provide a media blitz on special training events.

Print is the standard communication method available to the training officer. Training flyers are a common sight on the bulletin board, offering the basic information on a training program. This format works well for general training announcements and is the workhorse in the training announcement arsenal. It is a simple way to convey the necessary information and is extremely cost effective. Most can be created in any word processor software or can even be created by hand if no computer is available. Figure 8–4 (also provided in the cd-rom) offers a sample training announcement flyer that is common in the fire service.

Virginia Department of Fire Programs
2001 Hanover County Regional School

Held at Paramount's King's Dominion in the Paramount Theater – Doswell, Virginia

November 10 & 11, 2001
Hosted by:
Hanover Association of Volunteer Fire Companies

(Registration Begins at 08:00 a.m.; Opening Exercises begin at 08:30 a.m.)

"The Fire Service: Yesterday, Today & Tomorrow"

You can go to almost any fire station in Virginia and, sooner or later, someone will talk about how things used to be. Today, there seems to be a growing sense of frustration among the ranks as a generation who saw great developments in the fire service retires and the generation X'ers take over. "Whatever happened to the pride in the job?" or "What's in it for me?" they say. As the next generation of the fire service faces even more controversy and turmoil, it is vital that we get "back to the basics" to find the motivation and courage to face these future challenges.

The 2001 Hanover County Regional School will afford the members of the fire service the opportunity to discuss with the "Masters" the management ailments and frustrations within our fire service. The subject matter will be targeted to all members of the fire service, from the recruit firefighter to fire chief. If you have an issue with the fire service, here's your chance to have it addressed by the best.

(Scheduled to Appear)

"BOSS STUFF"
Alan V. Brunacini___ is a 43-year veteran of the Phoenix (AZ) Fire Department, where he has been Chief for 22 years. A past board chairman of the National Fire Protection Association, he currently heads the organization's Career Deployment Committee. He is the author of two books, **Fire Command** and **Essentials of the Fire Department Customer Service.**

"THE MILLENIUM ENGINE COMPANY"
Andrew A. Fredericks___, MIFireE, a 20-year veteran of the fire service, is a firefighter with Squad 18 of the Fire Department of New York. He is a New York state-certified fire instructor at the Rockland County Fire Training Center in Pomona, New York, and an adjunct instructor at the New York State Academy of Fire Science. He has two bachelor's degrees, one in political science and one in public safety, with a specialization in fire science, and a master's degree in fire protection management from John Jay College of Criminal Justice. He developed the **Fire Engineering** "Bread & Butter" Operations videos Advancing the Initial Attack Line (1998), and Methods of Structural Fire Attack (1999).

"PRIDE IN OWNERSHIP"
Rick Lasky___, a 22-year veteran of the fire service is Chief of the Lewisville (TX) Fire Department. Prior to Lewisville, he held the position of Fire Chief in Coeur D'Alene, Idaho and serviced as training officer with the Darien-Woodridge (IL) and Bedford Park (IL) Fire Departments. While in Illinois, he taught for the Illinois Fire Service Institute and Illinois Fire Chiefs' Association in a variety of programs and received the 1996 International Society of Fire Service Instructors "Innovator of the Year" award for his part in the development of the "Saving Our Own" program. He is an editorial advisor for **Fire Engineering** Magazine and serves on both the FDIC and FDIC WEST advisory committees.

Sunday afternoon will include an "open mic" session with all three presenters, offering a chance to present your questions to the panel.

Pre-Registration
Pre-registration is mandatory. All registration forms, along with fee, must be received by October 29, 2001. A registration fee of **$30** per student is mandatory and will include lunch and refreshments both days. Make checks payable to the Hanover Association of Volunteer Fire Companies. No telephone or fax registration will be accepted. For more information call (804) 371-0280.

Parking
Attendees should park in the employee parking lot (go to the base of the King's Dominion sign).
Do not go to the main gate.

Fig. 8–4 A Sample Training Announcement for a Special Event

In some cases, the training officer may wish to increase the promotion of a training class to achieve the best marketing mix. The next step up from a flyer is a promotional pamphlet. The pamphlet is a glossier promotional piece that provides detailed information about the training event. The training officer must consider whether an extra promotional push will benefit the program. Training that is specialized, new, or different may be a candidate for a pamphlet type promotion. Pamphlets can be produced using most basic publishing software and can even be done on word processors. They can range from one page, tri-fold pieces to multipage, catalog-type publications. A sample pamphlet is provided on the CD-ROM.

The Internet is a new marketing option for the training officer. Many departments are developing web sites that allow volunteers to access information at their leisure. The Internet also offers the ability to produce a much flashier promotional piece using animation and color. If the department has access to build a web site, the training officer must ensure that the training calendar is made available and constantly updated. Ensuring an up-to-date web site gives the reader confidence that the information they are receiving is up to the minute and will likely cause them to check the site periodically for new information.

Many departments use their radio systems to broadcast important announcements at a regularly scheduled time on a daily basis. This provides an opportunity to get training announcements on the air to the entire department at one time. Such announcements must be brief and should prompt the listener to seek out more information through other promotional sources. Making a radio announcement as a program is announced can help start the buzz about the training. Using a follow-up announcement just before the class will also help to motivate procrastinators to sign up.

No matter which promotional method is chosen, the training officer must ensure that the announcements are made well in advance. An attempt should be made to announce courses at a minimum of 30 days in advance. The further in advance volunteers have access to training information, the better prepared they will be to adjust their schedules. By incorporating consistency and marketing into the training regimen, the training officer will find improved success in getting volunteer firefighters into the classroom.

Place

Place is a commonly overlooked part of the marketing process when applied to volunteer firefighter training. Place refers to convenience and comfort and plays a vital role in the overall training experience. Many training officers assume that because their firehouse has only one meeting or training room, they are limited to what can be done to improve the place factor. Identify the most convenient and comfortable way to deliver the training, not based on what is available in the fire station, but on what is available in the community. Using the same methods to establish venues for special events, the training officer can tap into other resources that may make the training more appealing to the volunteer, or, if nothing else, offer an occasional change of scenery. Anything to keep the training program interesting and refreshing will help.

For volunteer departments with multiple stations, the training officer should consider moving the training programs around the various stations. This is in an effort to limit the amount of time and gas spent driving back and forth to training. On occasion, the training officer will find a concentration of students from one end of their jurisdiction. When this situation occurs, consider moving the class closer to the students, assuming there are no benefits to staying near specific training props or facilities.

Summary

Filling the shoes of the volunteer training officer is a challenging task. Using an organized approach when determining training needs and returning to the traditions of this great fire service will provide the training officer a solid foundation from which to work. The training officer must be a picture of the core values described in this text and should constantly strive to improve the training system. Think of those who have died in the line of duty and those who will make the ultimate sacrifice in the future to set the training pace for your department. Stand confident against those who feel volunteer firefighters shouldn't have to demonstrate high levels of performance, and most of all, look out for the brothers and sisters and make sure they are prepared when it is their time to risk it all.

Chapter Review Questions

1. What are the two main funding options for special events?

2. What are the main considerations for planning a special event?

3. Explain the Four P's of Marketing.

4. What are the main methods of communication available to the training officer when announcing training?

Answers

1. Full funding through the department budget, subsidies from outside organizations, or charging tuition or an admission fee.

2. Locating facilities, managing logistics, and demobilization.

3. The Four P's of Marketing are product, price, promotion, and place. The product refers to the variety of training available in the training system. Price describes the total impact on the volunteer including financial costs and time away from family. Promotion is used to communicate appealing factors about the training. Place refers to the convenience and comfort associated with the training program.

4. Print, Internet, and verbal communications are the main methods of communication available to the training officer.

Notes

[1] Kotler, Philip. 2000, *Marketing Management: The Millennium Edition*, Upper Saddle River, NJ: Prentice Hall Publishing. p. 3.

[2] Ibid., p. 15.

Index

V

W–Z

Prepare for the Fireground with Fire Engineering Books & Videos

FIREGROUND SIZE-UP

by Michael A. Terpak, Chief of the 2nd Battalion, Jersey City (NJ)

Fire officers have many decisions to make when they approach a scene—decisions that could mean the difference between life and death. Pre-incident information gives fire officers the advantage of knowing what to expect when they arrive at a fire. In this definitive guide to fireground size-up, Terpak gives firefighters an in-depth and expanded review of 15 size-up points to help them make decisions that are efficient, effective, and safe.

In each different type of building referenced, Terpak covers the following points:

- Construction concerns • Occupancy • Apparatus & staffing • Life hazard • Terrain • Water supply • Auxiliary appliances & aides • Street conditions • Weather • Exposures • Area • Location & extent of fire • Time • Height • Special considerations

Contents

The Fifteen Points of Size-Up • Private Dwellings • Multiple Dwellings • Taxpayers/Strip Malls and Stores • Garden Apartments and Townhouses • Row Frames and Brownstones • Churches • Factories, Lofts, and Warehouses • High-Rises • Vacant Buildings

407 pages/Hardcover/2002
ISBN 0-912212-99-3   **$74.95** US   **$ 89.95** INTL

FIREGROUND STRATEGIES

by Anthony Avillo, Deputy Chief, North Hudson (NJ) Fire and Rescue

This text is to be used as both a guide for the fireground strategist/tactician and the promotional candidate in preparing for a written exam. There are text and short answer questions as well as multiple choice scenarios, which are used by many testing authorities today. Each answer is explained in depth to help the reader understand the reason for the strategy or tactic presented. This text uses case studies extensively to drive points home. The text will allow the strategist to make decisions about such activities as line placement, ventilation considerations, and resource distribution, among other things. It will also allow the tactician to choose proper tactics in a given situation, enhancing the decision-making process on the fireground. It is the intent of this text, through diligent study and lesson reinforcement, to motivate, challenge, and strengthen the fireground strategist/tactician and/or the promotional candidate.

Contents:

Size-Up • Heat Transfer • Building Construction • Modes of Operations • Private Dwellings • Multiple Dwellings • High Rise • Contiguous Structures • Taxpayers and Strip Malls • Commercial Buildings • Hazardous Materials • Operational Safety

477 Pages/Hardcover/August 2002
ISBN 0-87814-840-X   **$59.95** US   **$74.95** INTL

FIREGROUND STRATEGIES SCENARIOS WORKB

Softcover/2003
ISBN 0-87814-840-X   **$54.95** US   **$69.95** INTL

3 Easy Ways
Online: www.penn
Phone: 1.800.752.9764
Fax: 1.877.218.1348